INTEGRATED PEST MANAGEMENT OF VEGETABLE CROPS

INTEGRATED PEST MANAGEMENT OF VEGETABLE CROPS

by

N. Emmanuel

A.V.D. Dorajee Rao

K. Gopal

L. Naram Naidu

M. Madhavi

T.S.S.K. Patro

2026

Daya Publishing House®

A Division of

Astral International Pvt. Ltd.

New Delhi – 110 002

ISBN: 978-93-5919-755-5

Published by : **Daya Publishing House®**
A Division of
Astral International Pvt. Ltd.
– ISO 9001:2015 Certified Company –
4736/23, Ansari Road, Darya Ganj
New Delhi-110 002
Ph. 011-43549197, 8130496929
E-mail: info@astralint.com
Website: www.astralint.com

Printed at : **Replika Press Pvt. Ltd.**

Dr. K. Gopal
Vice-Chancellor

Dr. Y.S.R. Horticultural University
Venkataramannagudem,
Tadepalligudem - 534101,
West Godavari District,
Andhra Pradesh
Ph: 08818-284312
Mobile: +91-7382633666
Email: vc@drysrhu.edu.in
Website: www.drysrhu.ap.gov.in

PREFACE

India is renowned as a prosperous nation characterized by diverse agro-climatic conditions, which significantly facilitates the cultivation of a wide array of horticultural crops. Currently, India ranks second only to China in terms of both the area and production of fruits and vegetables, contributing 10% of the global fruit output and 14% of the vegetable yield. India holds the second position in the worldwide production of onions and ranks third in the production of cabbage. Despite the presence of numerous advantageous factors, the anticipated growth in the horticultural sector has not been realized due to several limitations including pests and diseases.

Indiscriminate usage of pesticides has led to environmental degradation, adverse effects on human health, detrimental impacts on agricultural productivity and decline in agricultural sustainability. Promotion of Integrated Pest Management (IPM) practices is imperative at this juncture, as it represents an environmentally friendly strategy that incorporates all available alternative pest control methodologies such as cultural, physical, mechanical, and legal and biological control with greater emphasis on use of crop rotation, bio pesticides and plant origin pesticides.

Keeping this in view, the text book entitled "**INTEGRATED PEST MANAGEMENT OF VEGETABLE CROPS**" has been written and the publication deals with basics and applied aspects of IPM in vegetable crops. The chapters present an accurate and comprehensive review of IPM thus making the text an excellent coverage of practical guide to vegetable growers.

I accolade the authors for bringing out the Text book on *"INTEGRATED PEST MANAGEMENT OF VEGETABLE CROPS"*

(Dr. K. Gopal)

ABOUT THE AUTHORS

Dr. N. Emmanuel is a renowned Professor of Entomology at Dr. YSR Horticultural University, specializing in biological control, integrated pest management (IPM), toxicology, and the innovative use of drones and Artificial Intelligence (AI) for pest management. With over 80 research papers and textbooks to his name, he has earned prestigious accolades such as the Dr. M. S. Venugopal Gold Medal and the Dr. Guruprasad Pradhan Gold Medal. In 2023, he was honored with the Best Teacher Award by Dr. YSR Horticultural University. Dr. Emmanuel has led several impactful projects in coconut, cocoa, beekeeping, and vegetable crop IPM, significantly benefiting farmers. His commitment to advancing scientific beekeeping earned him the Excellence in Innovation Award in 2020 and the Scientist Award in 2021. Dr. N. Emmanuel composed and published Telugu video song "TENETEEGALU" in YouTube to raise awareness about the vital role of honey bees in horticulture, highlighting their importance in pollination and agricultural productivity.

Dr. A.V.D. Dorajee Rao, Associate Dean of Dr. YSRHU College of Horticulture, Venkataramannagudem, is a renowned academic and researcher with gold medals in B.Sc. (Ag.), M.Sc. (Ag.), and Ph.D. degrees. He has vast experience in teaching, research, and extension, publishing over 125 research articles and guiding numerous postgraduate and doctoral scholars. He was instrumental in establishing KVK Nellore and contributed significantly to national coconut research projects. With a proven track record in administration, he has held roles such as Warden, Officer-in-Charge for academics, and Campus Beautification Chairman.

Dr. K. Gopal, Vice-Chancellor of Dr. YSR Horticultural University, is a distinguished academician and researcher with 35 years of experience in teaching, research, and administration. He has held key roles such as Registrar, Controller of Examinations, and Zonal Research Head. A pioneer in crop protection, he developed groundnut varieties KRG 2 and KRG 3, disease-tolerant sesame variety Sweta Til, and canker-resistant acid lime varieties "Balaji" and "Vakula." Dr. Gopal established a Molecular Diagnosis Laboratory for virus-free citrus certification and developed Integrated Disease Management technologies. With 143 research publications and 124 conference papers, he has guided 6 Ph.D. and 27 PG students. His numerous awards include the Dr. Shyam Singh Best Scientist Award and Sardar Vallabhbhai Patel International Award. A Fellow of several scientific societies, his work has advanced agricultural and horticultural sciences in India.

Dr. L. Naram Naidu, Dean of Horticulture at Dr. YSR Horticultural University, A.P., is an alumnus of the Agricultural College, Bapatla under Acharya N.G. Ranga Agricultural University, Andhra Pradesh. With over 35 years in research, teaching, and extension, he has served in various academic and administrative positions, including Zonal Research Head, Director of Research, and Dean PG Studies. Developed seven chilli varieties, and contributed to improved spice crops. He played a key role in establishing research stations, securing external funding, and developing infrastructural facilities. Recognized for his contributions, he received the Best Seed Scientist award from the Seedsmen Association, Hyderabad, for his impact on the seed industry and farmers' service.

Dr. M. Madhavi earned her B.Sc. (Hort.) in 1987, M.Sc. (Hort.) in 1989, and Ph.D. (Hort.) in 2007. She is presently working as Director of Research at Dr YSR Horticultural University, venkatramannagudem, Andhra Pradesh. With over 10 years of research experience in citrus crop cultivation and 20 years in teaching, Dr. M. Madhavi is a renowned expert in the field of horticulture. She has guided 25 M.Sc. and 12 Ph.D. students and has published over 55 research papers, 20 popular articles, and 6 book chapters. She is also a member of various horticultural societies, showcasing her commitment to the field.

Dr. T.S.S.K. Patro, Principal Scientist at ARS, Vizianagaram, ANGRAU, is a distinguished plant pathologist specializing in small millet crops, with over 250 journal articles, 50 conference proceedings, 25 book chapters, and six patents to his name. He was conferred the prestigious Fakhruddin Ali Ahmed Award (2020) for groundbreaking research in tribal farming systems and the Best Institution Award (2022) under the Poshak Anaaj Awards. Dr. Patro's leadership earned his team the Best Performing AICRP Centre Award for small millets multiple times. He has also been honored with the Certificate of Special Recognition (2021) and the Sri Mandava Venkataramaiah Best Research Worker Award (2013) from ANGRAU. Additionally, he received the Acharya N.G. Ranga Young Scientist Award (2010) and several national and international awards.

CONTENTS

CHAPTER 1

PESTS OF BRINJAL

Brinjal, also known as eggplant or aubergine belongs to the family Solanaceae with diploid chromosome number 2n=24. It is one of the popular solanaceous vegetable crop grown all over the country and throughout the year for its tender fruits. Brinjal ranks fifth globally in economic value among vegetables, with top producers including China, India, Egypt, Turkey, and Indonesia.

In India, brinjal is referred by a range of names, notably Vashuthana (Malayalam), Kathiri (Tamil), Venkaya (Telugu), Begun (Bengali), Ringna (Gujarathi), Bhagan (Hindi), Badane (Kannada), Waangum (Kashmiri), Vange (Marathi), and Baigan (Oriya). India grows a wide variety of cultivars with varying fruit sizes, shapes, and colors. Brinjal is rich in nutrients but low in calories. Brinjal ranks fair in nutritional value in terms of carbohydrates, proteins, fibre and vitamins like Thiamin, Niacin, Pantothenic acid and Folacin as well as minerals like Calcium, Iron, Potash, Zinc, Copper and Manganese.

For centuries, brinjal is used in ayurvedic medicine in treating high blood pressure, diabetes, cuts, wounds, increases hunger, and lessens phlegm. It is rich in Vitamin B-6, fiber, potassium, and flavonoids, aiding in cholesterol reduction and blood pressure stabilization. Ideal for weight management and eggplants protect cell membranes for healthy brain function and memory. They also provide essential nutrients like calcium, magnesium, and phosphorus for strong bones.

India's brinjal area expanded by 0.29% to 6.77 lakh hectares, yielding 127.79 lakh tons in 2023–2024. The top states by acreage are Gujarat, Madhya Pradesh, Bihar, Chhattisgarh, Andhra Pradesh, West Bengal, Odisha, and Telangana. IARI, New Delhi, IIHR, Bangalore, and SAU's in India have released popular and noteworthy brinjal varieties like Pusa Purple Long, Pusa Purple Cluster, Pusa Kranti, Pusa Barsati, and more, like Manjri Gota, Vaishali, Arka Navneet, Arka Sheel, Arka Kusmukar, Arka Keshav (BWR - 21), Pusa Ankur, Gulabi, Bhagyamathi, and Shyamala.

Brinjal's average yield ranges from 20-30 t/ha depending on variety and season. Nevertheless, brinjal production and productivity is limited by insect pests like shoot & fruit borer, aphid, whitefly, and other pests which are discussed in this chapter along with their management practices.

1. Brinjal Fruit and Shoot borer , *Leucinodes orbonalis* (*Pyralidae, Lepidoptera*)

Identification: The adult moth is small with brown spots on its wings and is small, white with a pinkish and bluish hue. It lasts 6 to 10 days. Moths do not consume eggplant; instead, they feed on plant exudate or dew drops to stay alive. They are difficult to spot since they spend the daytime hiding under eggplant leaves. The moths emerge in the open to mate during the night. The abdomen of the female moth frequently curls forward. Adults live for about a week on average, with females outliving males.

Biology: The adult females lay eggs singly or in groups of two to five on the under surfaces of leaves, tender shoots, flower buds, or the base of developing fruits. Each female lays about 250 eggs, which are creamy white soon after laying, but turn red before hatching. The egg period is three to five days. The grown-up larva is pink with sparse hairs on the warts on the body and a dark brown or blackish head. Full-grown larva measures about 16-23 mm in length. Larva after hatching bores into the tender shoots and fruits and completes 6 larval instars in 16-22 days. Mature larva comes out and pupates outside the bore hole or in the stem. Pupa is present inside a dark brown tough silken cocoon. Pupal period varies from 9-14 days. The total life cycle is completed in 36-38 days on an average.

Damage: The larva is an internal feeder; as soon as it hatches, it bores into the closest tender stem, flower, or fruit. They quickly cover the entrance hole with excrement after burrowing into shoots or fruits. The afflicted twigs, flowers, and fruits subsequently dry out and may fall off. The immature shoot wilts as a result of larval feeding inside the shoot. The sight of withered shoots is the most visible sign that this insect has damaged an eggplant field. Eventually, the damaged shoots wither and stop growing. As a result, the number and size of fruits are decreased along with plant growth. Although new shoots can appear, this slows crop maturation and leaves the newly developed shoots vulnerable to damage.

Management :

1. Avoid monoculture and follow crop rotation.
2. Avoid growing eggplant seedlings near fields with standing crops, in or near fields where the crop was grown previously, or near dried eggplant heaps. If seedlings must be grown in those areas, cover the beds with 30-mesh nylon net to prevent the entry of FSB moths.
3. Choose resistant or moderately resistant cultivars available in the region. For instance, accessions or varieties such as EG058, Pusa Purple Long, Pusa Purple Cluster, Pusa Purple Round, H- 128, H-129, Aushey, Thorn Pendy, Black Pendy, H-165, H-407, Dorley, PPC-17-4, PVR-195, Shyamla Dhepa,

Banaras Long Purple, Arka Kesav, Arka Kusmakar, Punjab Barsati, Punjab Chamkila, Kalyanpur-2 and Gote-2 have been reported to be tolerant or resistant

4. Remove and destroy infested shoots and fruit at regular intervals until final harvest.
5. Weekly releases of egg parasitoid, *Trichogramma chilonis* Ishii @ 1g parasitized eggs/ha/week and larval parasitoid, *Bracon habetor* Say @ 800-1000 adults/ha/week could be followed .
6. Install FSB sex pheromone lures in traps at the rate of 10 traps per hectare.
7. Spray Azadirachtin 1.0 % EC (10000 ppm) @2.0 ml/ lit., Chlorantraniliprole 18.5 SC @ 60 ml/ha, Emamectin benzoate 5 % SG @4 g/10 lit , Flubendiamide 20 WG@5 g/10 lit., Novaluron 10 % EC @7.5 ml/10 lit., Phosalone 35 % EC @13 ml/10 lit.

2. Brinjal Stem borer, Euzophera perticella (Pyralidae, Lepidoptera)

Identification : Moth is medium sized and pale in colour. The forewing is pale yellow or grayish- brown in colour, with black lines in the middle and the hind wings are white. The larva is white or yellowish white in colour with several bristly hairs and an orange- brown or red head.

Biology : The cream- coloured eggs are laid either singly or in groups on the tender leaves, shoots and petioles. The eggs are elongate and flat. A single female lays about 150- 200 eggs. The egg period varies from three to ten days. Eggs hatch in 3-5 days and young larvae bore into the stem and feed on the pith making longitudinal tunnels. They make silken cocoon within the feeding galleries and pupates in 7- 10 days. They transform into adults in 6-8 days.

Damage: Soon after hatching, the larva starts boring into the stem near ground level. Mostly they bore in the branching area or in the leaf axils and seal the entry holes with excretory materials. Larvae feed downward along the length of the main stem, which results in stunted growth or wilting and withering of the whole plant. The later stages of plant growth are most vulnerable to this insect.

Management:

1. Remove and promptly destroy the infested plants.
2. Avoid ratoon cropping.
3. Release of parasitoid Pristomerus euzopherae.
4. Install Light trap @ 1/ha to attract and kill adults.
5. Spray neem oil 2ml/lit ,Imidacloprid 80.5 SC 0.6 ml/lit, Chlorpyriphos 20 EC 2ml/lit, Thiamethoxam 25 WSG 0.6 mg/lit, Profenophos 2ml/lit.
6. Avoid using synthetic pyrethriods causing resurgence.

3. Spotted leaf beetle (or Hadda Beetle), *Henosepilachna vigintioctopunctata* (Fabr.) (Coccinellidae: Coleoptera)

Identification: The subfamily Epilachninae contains plant-feeding ladybird beetles because most other ladybird beetles are predators, not plant pests. These brownish or orange colored, hemispherical beetles are larger than other ladybird species. *H. vigintioctopunctata* (in Latin, *viginti* means 20 and *octo* means 8) has 28 black spots on the forewing (elytra). *E. dodecastigma* (*dodeca* means 12 in Greek) has 12 black spots on the elytra. However, beetles with 14, 16, 18, 20, 22, 24 or 26 spots have been observed under field conditions, due to mating between females of *E. dodecastigma* and males of *H. vigintioctopunctata* .

Biology: The female lays elongate, spindle-shaped yellowish eggs in groups of 10 - 20 on the under surface of leaves. About 120 - 180 eggs may be laid by a female. The egg period is 2 - 4 days. The yellowish spiny grubs become full grown in 10 -35 days and pupate on the leaf or stem. The pupa is hemispherical, yellowish with spines on the posterior part. The anterior portion being devoid of spines. Adults emerge in a week and live for a month feeding on leaves. The total life-history takes 17 - 50 days depending on weather conditions.

Damage: The grub and adult have chewing mouthparts. Hence, they scrape the chlorophyll from the epidermal layers of the leaves. The feeding results in a typical ladder-like window. The windows will dry and drop off, leaving holes in the leaves. In severe infestations, several windows coalesce together and lead to skeletonization and the formation of a papery structure on the leaf.

Management:

1. Choose resistant or moderately resistant cultivars available in the region. Varieties such as Arka Shirish, Hissar Selection 14, and Shankar Vijay have been reported to be tolerant or resistant to Epilachna beetle, especially *E. vigintioctopunctata.*
2. Conserve the population of parasitoids such as *Pediobius foveolatus.*
3. Collection and destruction of affected leaves along with the eggs, grubs and adults.
4. Spray application of Chlorantraniliprole 18.5% SC @ 0.4 ml/lt , Spinetoram 11.7% @ 6.5 ml/10 lit SC @ 0.25 ml/lt, Indoxacarb 14.5% SC, Chlorpriphos 20 EC 2ml/lit,, profenofos 0.05% or DDVP @ 1 ml/lit or Neem oil @ 3 ml/ lit.

4. Brinjal Lace- wing bug, *Urentius histricellus* (Tingidae, Hemiptera)

Identification: Adults are straw coloured on dorsal side and black coloured on ventral side with sculptured forewing and pronotum. Nymphs are yellowish white with prominent spines. Eggs are white nibble shaped.

Damage: Infestation starts after few weeks of transplantation. The nymphs bore into the growing shoots or petioles large leaves and feed on internal tissues. As a result of damage, affected shoots wither and plants exhibit the symptoms of drooping. After fruit formation larva make their entry under the calyx when they are young. The holes later plugged with excreta leaving no visible sign of infestation. Large holes seen on the fruits are the exit holes.

Management: Spray methyl dematon 25 EC @ 1ml/lit, dimethoate 30 EC @ 1ml/lit.

5. Leafhopper, Amrasca devastans Distant (Hemiptera: Cicadellidae)

Identification: The nymphs resemble the adults, but lack wings. Instead, they have slightly extended wing pads. They are pale green in colour. They tend to move sideways when disturbed. Adults are wedge- shaped, pale green insects. They have fully developed wings with prominent black spots on each forewing.

Damage: Both nymphs and adults suck the sap from the lower leaf surfaces through their piercing and sucking mouthparts. While sucking the plant sap, they also inject toxic saliva into the plant tissues, which leads to yellowing. When several insects suck the sap from the same leaf, yellow spots appear on the leaves, followed by crinkling, curling, bronzing and drying or "hopper burn". Leafhoppers also damage in okra, cotton and potato.

Management:

1. Choose tolerant or resistant cultivars with hairy leaves, as the length and density of the trichomes (hairs) repel leafhoppers such as Manjari Gota, Vaishali, Mukta Kesi, Round Green, and Kalyanipur T3 and Bangladeshi variety. Bagun 6 are reported to be less susceptible or tolerant to damage.
2. Monitor Leafhoppers with yellow (570-580 nm) sticky traps .
3. Grow okra as a trap crop along the borders of an eggplant field; if pesticides are required to control an infestation, they can be applied to the trap crop.
4. Generalist predators such as ladybird beetles and green lacewings and Parasitoids such as *Anagrus flaveolus* Waterhouse and *Stethynium triclavatum* are effective against leafhopper.
5. Spray Neem seed kernel extract (NSKE) @ 5%, Acetamiprid 20% SP (10 g/10 L), Thiamethoxam 25% WG (3g/10 L) ,Imidacloprid 75% WG (1.5 g/10 L).

6. Whitefly, *Bemisia tabaci* (Hemiptera: Aleyrodidae)

Identification: The whitefly is slightly smaller (about 0.96 mm in the female and 0.82 mm in the male) and slightly yellower than most other whitefly pests .The wings are covered with powdery wax and the body is light yellow in colour. The head is broad at the antennae and narrow towards the mouth parts. The wings are held roof-like at about a 45 angle, whereas other whiteflies usually hold the wings nearly flat over the body. Hence, appears more slender than other common whiteflies. The first instar nymph has antennae, eyes, and three pairs of well developed legs. The nymphs are flattened, oval-shaped, and greenish-yellow in colour. The legs and antennae are atrophied during the next three instars and they are immobile during the remaining nymphal stages. The last nymphal stage has red eyes . This stage is sometimes referred to puparium, although insects of this order (Hemiptera) do not have a perfect pupal stage (incomplete metamorphosis).

Biology: The females mostly lay eggs near the veins on the underside of leaves. They prefer hairy leaf surfaces to lay more eggs. Each female can lay about 300 eggs in its lifetime. Eggs are small (about 0.25 mm), pear-shaped, and vertically attached to the leaf surface through a pedicel. Newly laid eggs are white and later turn brown. The eggs are not visible to the naked eye, and must be observed under a magnifying lens or microscope. Egg period is about three to five days during summer and 5 to 33 days in winter . Upon hatching, the first instar larva (nymph) moves on the leaf surface to locate a suitable feeding site. Hence, it is commonly known as a "crawler." It then inserts its piercing and sucking mouthpart and begins sucking the plant sap from the phloem. Nymphal period is about 9 to 14 days during summer and 17 to 73 days in winter . Adults emerge from puparia through a T-shaped slit, leaving behind empty pupal cases or exuviae. Adults live from one to three weeks.

Damage: Both the adults and nymphs suck the plant sap and reduce the vigour of the plant. In severe infestations, the leaves turn yellow and drop off. When the populations are high they secrete large quantities of honeydew, which favours the growth of sooty mould on leaf surfaces and reduces the photosynthetic efficiency of the plants.

Management:

1. The field selected for eggplant or seedling production should be clean and not be located near any host plants and weeds.
2. Grow eggplant seedlings in insect-proof (50–64 mesh) net houses, net tunnels, greenhouses, or plastic houses.
3. If the seedlings are produced under open field conditions, use yellow sticky traps at the rate of 1-2 traps/50-100 m2 to trap the whiteflies. Hang the traps slightly above or at the canopy level for better trapping.
4. Maintain a high standard of weed control in seedling production areas and crop fields to reduce the availability of alternate host plants.
5. Plant fast-growing crops like maize, sorghum, or pearl millet in the border of the field to act as barriers to reduce whitefly infestations.
6. Reflective plastic or straw mulches may reduce landing of whiteflies on eggplant crops.

7. Spray with 0.05% dimethoate or 0.01% imidacloprid or 0.1% acetamiprid
8. In case of severe infestation, two sprayings at 10 - 12 days interval with 0.03% oxydemeton methyl or Emamectin benzoate 5 % SG@ 4 g/10 lit, Fipronil 5 % SC@1.5 ml/lit., Spinosad 45 % SC @3.2 ml/10 lit.

7. Aphids, Aphis gossypii (Aphididae, Hemiptera)

Biology: When first deposited, the eggs are yellow, but they soon become shiny black in colour. The nymphs vary in colour from tan to gray or green, and often are marked with dark head, thorax and wing pads, and with the distal portion of the abdomen dark green. The body is dull in color because it is dusted with wax secretions. The nymphal period averages about seven days. The wingless (apterous) parthenogenetic females are 1 to 2 mm in length. The body is quite variable in color: light green mottled with dark green is most common, but also occurring are whitish, yellow, pale green, and dark green forms. The legs are pale with the tips of the tibiae and tarsi black. The cornicles also are black. Small yellow forms apparently are produced in response to crowding or plant stress. Winged (alate) parthenogenetic females measure 1.1 to 1.7 mm in length. The head and thorax are black, and the abdomen yellowish green except for the tip of the abdomen, which is darker. The wing veins are brown. The egg-laying (oviparous) female is dark purplish green; the male is similar. The duration of the adult's reproductive period is about 15 days, and the post-reproductive period five days. These values vary considerably, mostly as a function of temperature. The optimal temperature for reproduction is reported to be about 21 to 27 degrees. C. *Viviparous* females produce a total of about 70 to 80 offspring at a rate of 4.3 per day.

Damage: Although A. *gossypii* is polyphagous, it prefers to feed on cotton and cucurbit vegetables; it is commonly known as "cotton aphid" or "melon aphid." Both the nymphs and adults possess piercing and sucking mouthparts. They occur in large numbers on the tender shoots and lower leaf surfaces, and suck the plant sap. Slightly infested leaves exhibit yellowing. Severe aphid infestations cause young leaves to curl and become deformed. Like whitefly, aphids also produce honeydew, which leads to the development of sooty mould.

Management:

1. Although A. *gossypii* is a polyphagous insect, it overwhelmingly prefers to feed on cucurbits and cotton. Hence, the field selected for egg plant or seedlings production should be located away from cucurbits and cotton.
2. Grow eggplant seedlings in insect-proof (50–64 mesh) net houses, net tunnels, greenhouses, or plastic houses to avoid early infestation.
3. The ladybird beetles (*Menochilus* sp. and *Coccinella* sp.) and green lacewings are efficient predators of aphids. Protect the population of these predators by avoiding the use of broadspectrum pesticides. Inundative release of ladybird beetles @ 200 pairs per ha at fortnightly intervals can suppress the aphid population.

4. *A.gossypii* can develop resistance to pesticides. Use only those pesticides that have been recommended.
5. Do not use the same compound or pesticide group continuously to avoid the development of pesticide resistance in insects.
6. Imidacloprid 17.8 % SL @3.0 ml/10 lit, Dimethoate 30 % EC @1.0 ml/lit., Emamectin benzoate 5 % SG@ 4 g/10 lit, Fipronil 5 % SC@1.5 ml/lit., Spinosad 45 % SC @3.2 ml/10 lit, Thiacloprid 21.7 % SC @6.0 ml/10 lit. Ethion 50 % EC @2.0 ml/lit., Oxydemeton –Methyl 25 % EC @1.0 ml/lit, Phosalone 35 % EC@2.0 ml/lit.

8. Cestius (Hishimonus) phycitis (Cicadellidae, Hemiptera)

Identification: Adults are small light brown leafhopper, measuring around 3 mm long. Males are little smaller in body length. Nymphs are creamy white, wingless and turn brownish with the advancement of age. They are found between the veins of leaves on the undersurface.

Damage: Reduction in leaf size and rosette appearance are the most prominent symptoms of little leaf disease of brinjal or eggplant. This disease is transmitted by a leafhopper (*Hishimonus phycitis*). In severe cases, affected plants do not bear any fruit, or, if formed, it becomes hard and tough.

Management:

Spray application of Dimethoate 30 EC @ 2 ml/lit. or Methyl dematon 25 EC @ 2 ml/lit. or fipronil @ 2 ml/lit. with a waiting period of 15 days.

9. Blister beetle, *Mylabris pustulata* Thunberg (Coleoptera : Meloidae)

Identification: The adult **Mylabris pustulata** is about 2.0-2.5 cm in length and bears red or reddish orange and black alternating bands on the forewing (elytra),Adults are rather soft-bodied, long-legged beetles with the head deflexed, fully exposed, and abruptly constricted behind to form an unusually narrow neck, the pronotum much narrower at the anterior end than the posterior and not carinate laterally, the forecoxal cavities open behind, and each of the tarsal claws cleft into two blades. Body length generally ranges between 3/4 and 2 cm.

Biology: Eggs are laid in masses in the ground/hard soil in 20 batches. Females dig soil and lay 60 eggs at the same time of grass hopper. Incubation period is 20 – 22 days. Newly hatched larva is called as TRIANGULIN, very active on egg pods of grass hoppers. Larval Period is 10 – 30 days. Each larvae feeds upto 40 – 70 GH eggs. Pupation in soil upto 2 months. Female lays 10,000 to 20,000 eggs during life span and Over winters in egg stage.

Damage: The adult is the destructive stage. As the insects feed on the plants' reproductive parts, flowers they can cause significant yield losses.

Management:

1. Pick off beetles by hand (wear gloves or use insect nets) and destroy.
2. Spray during flowering stage of the crop with any of the following insecticides viz., Cypermethrin or Fenvelerate or Profenophos or Acephate

10. Red spider mite, *Tetranychus utricae* Koch (Acarina : Tetranychidae)

Identification: *T. urticae* is commonly known as red spider mite or twospotted spider mite. They are minute in size, and vary in color (green, greenish yellow, brown, or orange red) with two dark spots on the body.

Biology: Eggs are round, white, or cream-colored egg period is two to four days. Eggs are laid singly, up to 100 per female, during her 3- to 4-week life span. Eggs hatch into larvae in as few as 3 days. Following a brief larval stage, several nymphal stages occur before adults appear. Egg to adult cycle can be completed in 7-14 days depending on temperature. There are several overlapping generations in a year. The adult lives up to three or four weeks.

Damage: Spider mites usually extract the cell contents from the leaves using their long, needle- like mouthparts. This results in reduced chlorophyll content in the leaves, leading to the formation of white or yellow speckles on the leaves . .Under high population densities, the mites move to the tip of the leaf or top of the plant and congregate White and yellow speckles caused by spider mites . Webbing of leaves by spider mites using strands of silk to form a ball-like mass, which will be blown by winds to new leaves or plants, in a process known as "ballooning."

Management:

1. Predatory mites such as *Phytoseiulus persimilis* and several species of *Amblyseius,* are effective under protective structures and in high humidity.
2. *Mallada basalis* Walker and *Chrysoperla carnea* also are effective generalist predators of spider mites.
3. Avoid weedy fields and do not plant eggplant adjacent to legume forage crops.
4. Avoid early season, broad-spectrum insecticide applications for other pests.
5. Do not over-fertilize.
6. Overhead irrigation or prolonged periods of rain can help reduce populations.
7. For chemical control, use selective products whenever possible. Selective products which have worked well in the field include: bifenazate (Acramite), abamectin (Agri-Mek), spirotetramat (Movento), spiromesifen (Oberon 2SC): Group 23, mainly affects immature stages.

CHAPTER 2

PESTS OF TOMATO

Tomato (*Solanum lycopersicum* L.) is one of the most important vegetable crops in the world. It belongs to the family Solanaceae with the chromosome number of 2n=2x=24. Tomatoes are grown globally in various climates. They are pubescent herbs with curled pinnate leaves, white flowers, and red or yellow fruits. It is a self pollinated crop. The major tomato producing states are Maharashtra, Bihar, Karnataka, Uttar Pradesh, Orissa, Andhra Pradesh, Madhya Pradesh and Assam.

Tomatoes are nutritious and popular worldwide for their antioxidants and essential nutrients. They are rich in vitamins, minerals, and carotenoids like lycopene, known for its benefits such as cancer prevention, heart health, and improved cognitive function. Consuming tomatoes can increase antioxidants in the body and improve overall health. Over 15,000 tomato varieties exist globally, with over 1,000 grown in India. While only a few are sold commercially, where IARI, IIHR, SAU's and seed companies continue to develop new varieties with special qualities namely, Arka Abhijith, Abhinav,Namdhari, Rashmi, Vaishali, Rupali,Arka Shrestha, Arka Vardan, Arka Vishal, Arka Vikas, Pusa Early Dwarf , Pusa Ruby, Pusa Gaurav.

Tomato is a key vegetable in India, accounting for 11.04% of total vegetable production. It is known as a "protective food" for its nutrition but is limited by perishability. Top global producers include China, India, Turkey, USA, Italy, Egypt, and Spain. In India, tomato cultivation area in 2022-23 was 8.49 lakh ha (20.97 lakh acres), producing 204.25 Lakh Tonnes. Major states include Madhya Pradesh, Odisha, Karnataka, West Bengal, Andhra Pradesh, Gujarat, and Tamil Nadu.

Tomato production has been intensified over the years, however, yields continue to be low due to several production constraints such as pests, diseases and environmental factors. The incidence of insect-pests may vary from season to season and crop growth stages. The major pests attacking tomatoes are fruit borer, whitefly, thrips, red spider mites, leaf miners, cutworms and aphids. Tomato fruit borer, *Helicoverpa armigera* is an important pest which causes considerable losses in quantity as well as quality of tomato fruits. The monetary loss due to this pest in India has been estimated over rupees one thousand corer per year and yield losses ranged from 14 – 100 percent on different crops. Whitefly and aphids are the most serious sucking pests, which have been reported to cause about 45%and 34% yield loss in tomato, respectively.

1. Tomato fruit borer, *Helicoverpa armigera* (Noctuidae: Lepidoptera)

Identification: Moths are medium sized, stout, ochreous with olive grey forewings in male and reddish brown forewings in female with dark brown circular spot in the centre and indistinct double waved antemedial lines. Hind wings are pale smoky-white with a broad blackish outer border.

Biology: The eggs are yellowish-white, ribbed and dome-shaped, 0.4 - 0.5 mm in diameter. Freshly hatched larvae are yellowish-white in colour but gradually change and acquire greenish tinge. Full grown caterpillars are 40 - 48 mm long, apple-green in colour with whitish and dark-grey broken longitudinal stripes. Full grown caterpillars drop down from the plants and burrow in the soil where they pupate. Pupae are dark-brown in colour, 11 - 14 mm long and have a sharp spine at the anal end. Fecundity of females is rather high ranging between 1200 - 1600 eggs. Incubation, larval and pupal stages last for 2 - 4, 15 - 24 and 10 - 14 days respectively and the entire life-cycle may be completed in 4 - 6 weeks.

Damage: Eggs are laid singly, generally on leaves and flowers but sometimes on fruits as well. On hatching, the young larvae feed on tender foliage; advanced stage larvae attack the fruits. They bore circular holes and thrust only a part of their body inside the fruit and eat the inner contents. If the fruit is bigger in size, it is only partly damaged by the caterpillar but later it is invariably invaded by fungi and bacteria and spoiled completely. The larvae move from one fruit to another and a single caterpillar may eat and destroy 2 - 8 fruits.

Management:

1. Hand-picking of caterpillars and their mechanical destruction in the early stage of infestation can keep the population of this pest under check.
2. Spray application of HaNPV@ 250 larval equivalents/ha in 1125 lit. of water/ha when the larval population reaches 1 larva / m^2. (or) *Bacillus thuringiensis* var. *kurstaki* @ 1.5 kg in 1125 lit. of water/ha. at evening hours.
3. Installation of pheromone traps @5/ha coinciding with the initiation of flowering and destruction of collected moths.
4. Release of the egg parasitoid, *Trichogramma chilonis* or *T. brasiliensis* @ 1Lakh/ha coinciding with flower initiation at 15 days interval.
5. Planting of marigold as a trap crop (1:16 ratio). Marigold is to be planted 20 days before tomato planting
6. Erection of 20 bird perches /acre

7. Spray Azadirachtin 1.0 % EC (10000 ppm) @2.0 ml/ lit.,Indoxacarb 14.5 % SC @8 ml/10 lit. , Flubendiamide 20 WG@5 g/10 lit., Novaluron 10 % EC @7.5 ml/10 lit.

2. Whitefly, *Bemisia tabaci* (Aleyrodidae : Hemiptera)

Identification: Adults are minute insects, about one mm long, covered completely with a white waxy bloom. Nymphs are oval, scale- like and greenish - white in colour.

Biology: Egs are pear- shaped, light yellowish in colour, about 2mm long and can be seen standing upright on leaves, being anchored by a tail- like appendage inserted into the stoma of leaves. On hatching, the nymphs crawl a little, settle down on a succulent spot on the same leaf and never move again during that stage. Incubation period 3 – 5 days in summer extending upto 33 days during winter. Nymphal development takes 9 – 14 and 17 – 81 days in summer and winter, respectively and pupal period lasts for 2 – 8 days being longer during winter than in summer. A life – cycle may be completed in as little as 14 days or it may even be prolonged up to 107 days. There are about 12 overlapping generations in a year.

Damage: White, tiny, scale – like insects may be seen darting about near the plants or crowding in between the veins on ventral surface of leaves, sucking the sap from the infected parts. The pest is more active during the dry season and its activity decreases with the onset of rains. As a result of their feeding the affected parts become yellowish, the leaves wrinkle and curl downwards and are ultimately shed. Besides the feeding damage, these insects also excude honeydew which favours the development of sooty mould. In case of severe infestation, this black coating is so heavy that it interferes with the photosynthetic activity of the plant resulting in stunted growth. This whitefly also acts as a vector, transmitting the leaf curl virus.

Management:

1. The field selected for tomato or seedling production should be clean and not be located near any host plants and weeds.
2. If the seedlings are produced under open field conditions, use yellow sticky traps at the rate of 1-2 traps/50-100 m2 to trap the whiteflies.
3. Install the traps slightly above or at the canopy level for better trapping.
4. Maintain a high standard of weed control in seedling production areas and crop fields to reduce the availability of alternate host plants.
5. Plant fast-growing crops like maize, sorghum, or pearl millet in the border of the field to act as barriers to reduce whitefly infestations.
6. Reflective plastic or straw mulches may reduce landing of whiteflies on tomato crops.

7. Spraying with 0.05% dimethoate or 0.01% imidacloprid or 0.1% acetamiprid. In case of severe infestation, two sprayings at 10 - 12 days interval with 0.03% oxydemeton methyl or Spinosad 45 % SC @3.2 ml/10 lit.

3. White Tailed Mealy Bug, *Ferrisia virgate* (Cockerell) (Coccidae : Homoptera)

Identification: Eggs of this mealy bug are pale- yellow, cylindrical and about 0.3 mm long. Freshly hatched crawlers are yellowish in colour and become pale white in 2 - 3 days. Adult females are apterous, long, slender, slightly oval (3.5 – 4.5 x 1.5 – 2.0 mm) covered with dusty white waxy secretion and having a pair of conspicuous long glossy wax tassels at the caudal end.

Biology: A single female lays 100 - 400 eggs which remain concealed under the female. Reproduction is sexual as well as parthenogenetic. Incubation period is15 minutes to 4 hours and the immature stages may last for about 20 - 60 days in case of male and 19 - 47 days in case of females. Longevity of males is 1 - 3 days while the females live for 5 - 7 weeks.

Damage: Eggs are laid in clusters in cottony ovisac which remains concealed under the female. On hatching, the crawlers remain huddled together in cottony nest under the body of the mother. Later, these crawlers become active and wander about, moving swiftly till they find a succulent spot where they puncture the epidermis, inject their toxic saliva and start sucking the cell sap. The mechanical injury thus caused also serves as an entry for various disease producing organisms (bacteria and fungi). From 2nd instar onwards the nymphs secrete honeydew on which black sooty mould develop, which in turn hinders the photosynthetic activity of the plant resulting in stunted growth.

Management:

1. Early detection provides the best chance at managing mealy bugs due to the vulnerability of the immature stages.
2. Remove and destroy mechanically all the affected leaves and twigs in the early stages of infestation.
3. Mealybug destroyer (*Cryptolaemus montrouzieri*), green lacewings and brown lacewings can be released to prevent mealybug outbreaks.
4. Spray application of Imidacloprid 17.8 SL 0.3 ml/lit, Chlorpyriphos 20 EC 2ml/lit, Thiamethoxam 25 WSG 0.6 mg/lit, Profenophos 2ml/lit.

4. Serpentine leaf miner, *LIriomyza trifolii* (Agromyzidae, Diptera)

Identification:

Egg: Eggs tend to be deposited in the middle of the plant; the adult seems to avoid immature leaves. The female deposits the eggs on the lower surface of the leaf, but they are inserted just below the epidermis. Eggs are oval in shape and small in size, measuring about 1.0 mm long and 0.2 mm wide. Initially they are clear but soon become creamy white in color.

Larva: Body and mouth part size can be used to differentiate instars; the latter is particularly useful. For the first instar the mean and range of body and mouth parts (cephalopharyngeal skeleton) lengths are 0.39 (0.33 to 0.53) mm and 0.10 (0.08 to 0.11) mm, respectively. For the second instar the body and mouth parts measurements are 1.00 (0.55 to 1.21) mm and 0.17 (0.15 to 0.18) mm, respectively. For the third instar the body and mouth parts measurements are 1.99 (1.26 to 2.62) mm and 0.25 (0.22 to 0.31) mm, respectively. A fourth instar occurs between puparium formation and pupation, but this is a nonfeeding stage and is usually ignored by authors. The puparium is initially golden brown in color, but turns darker brown with time.

Adult : Adults are small, measuring less than 2 mm in length, with a wing length of 1.25 to 1.9 mm. The head is yellow with red eyes. The thorax and abdomen are mostly gray and black although the ventral surface and legs are yellow. The wings are transparent. Key characters that serve to differentiate this species from the vegetable, **Liriomyza sativae** Blanchard, are the matte, grayish black mesonotum and the yellow hind margins of the eyes. In vegetable leafminer the mesonotum is shining black and the hind margin of the eyes is black. The small size of this species serves to distinguish it from pea leafminer, **Liriomyza huidobrensis** (Blanchard), which has a wing length of 1.7 to 2.25 mm. Also, the yellow femora of American serpentine leafminer help 2.26 to separate it from pea leafminer, which has darker femora.

Biology: Leafminers have a relatively short life cycle. The time required for a complete life cycle in warm environments is often 21 to 28 days, so numerous generations can occur annually in tropical climates. At 25 C the egg stage required 2.7 days for development; the three active larval instars required 1.4, 1.4, and 1.8 days, respectively; and the time spent in the puparium was 9.3 days. Also, there was an adult preovipostion period that averaged 1.3 days. The temperature threshold for development of the various stages is 6 to 10 C except that egg laying requires about 12 C.

Damage: Punctures caused by females during the feeding and oviposition processes can result in a stippled appearance on foliage, especially at the leaf tip and along the leaf margins. However, the major form of damage is the mining of leaves by larvae, which results in destruction of leaf mesophyll. The mine becomes noticeable about three to four days after oviposition and becomes larger in size as the larva matures. The pattern of mining is irregular. Both leaf mining and stippling can greatly depress the level of photosynthesis in the plant. Extensive mining also causes premature leaf drop, which can result in lack of shading and sun scalding of fruit. Wounding of the foliage also allows entry of bacterial and fungal diseases.

Management:

a. Release of parasitoids *Diglyphus isaea* (Walker) (Hymenoptera: Eulophidae) and *Dacnusa sibirica* Telenga (Hymenoptera: Braconidae) and entomopathogenic nematode *Steinernema carpocapsae*

b. Spray application :

   1. *Isaria Fumosorosea* (*Hypocreales: Cordycipitaceae*)
   2. neem based formulations @ 3 ml/lit, NSKE 5%
   3. Acephate @ 1ml/lit.
   4. profenophos @ 2 ml/lit,
   5. Cyantraniliprole 10% OD @ 90 and 105 g a.i./ha,
   6. fipronil @ 1ml/l ,
   7. diafethiuron @ 2g/l ,
   8. spinosad @ 0.25 ml/l ,
   9. imidachloprid @ 0.3 ml /l

CHAPTER 3

PESTS OF BHENDI

Okra or Lady's finger [*Abelomoschus esculentus* (L.) Moench], belongs to the Malvaceae family and is an amphidiploid plant with chromosomal number 2n = 130–140. The vast majority okra is grown in tropical and subtropical climates. When it comes to the area under vegetables grown in the nation, it is ranked fifth, next to tomatoes. India leads the globe in okra production, accounting for 60% of the global market followed by Nigeria and Mali.

Bhendi a common vegetable grown in India is referred by a range of names, notably Ladies Finger, Okra, Bhindi (Hindi), Dhenras (bengali), Vendai (Tamil), bhindo (Gujarati), Bendekayi (Kannada), Ventaykka (Malayalam), Asra-pattraka (Sanskrit), etc. India produces various okra cultivars with diverse sizes, shapes, and colors. It is harvested for its young tender fruits, commonly used in cooking. Okra is a nutritious vegetable, rich in vitamins, minerals, and protein. It is low in calories but high in vitamin C and vitamin K, essential for immune function and blood clotting. Additionally, okra has high water content, dietary fiber, and carbohydrates, along with good amounts of minerals, thiamin, folate, and magnesium.

Lady finger contains fiber which can help reduce appetite and promote feelings of fullness. It also has medicinal properties that may lower bad cholesterol, total cholesterol, and triglyceride levels. Additionally, it is a good source of iodine and can be used to treat goiter. The fruit can be dried or frozen for off-season use, while the skin and fibers are used in paper and cardboard production. The root and stem are used to clear cane juice for making jaggery.

India is the largest producer of okra in the world with a production of 33.24 lakh metric tonnes from an area of 3.47 lakh hectares. The major bhendi (okra) producing states are Uttar Pradesh, Bihar, West Bengal, Odisha, Assam, Andhra Pradesh and Karnataka. India has released many remarkable varieties like Pusa Makhmali, Pusa Sawani, ASel-10 (Arka Anamika), Sel-4 (A. Abhay), -7, Punjab Padmini, Punjab no. 13, Punjab 8, DVR 1, 2, 3, 4, VRO 3, 4, 6, 5, 22, 25, Parbhani

Kranti, Kiran, Salkeerthi, Co 1, Aruna, MDU 1, Varsha Uphar, Azad Kranthi and Hybrids : CO 2, 3, COBhH 1from IARI, IIHR, PAU and IIVR .

Bhendi yield ranges from 7.5-10 t/ha, with hybrids yielding 15-22 t/ha. Insect pests like shoot and fruit borer limit production and productivity, along with other sucking pests are discussed in this chapter.

1. Okra Shoot and Fruit borer, *Earias vittella* & *E. insulana* (Arctidae: Lepidoptera)

Identification:

E. vittella : Eggs are spherical in shape, about half mm in diameter, light bluish-green in colour and beautifully sculptured with 26 - 32 longitudinal ridges; the alternate ridges project upwards to form a crown. Adults Fore wings buff coloured with a green wedge.

E. insulana : Larva: Brown coloured with dorsum showing a white median longitudinal streak. All the abdominal segments have two pairs of fleshy tubercles (one dorsal and other lateral). Adults has completely green fore wings.

Biology: A female lays on an average 400 eggs (65 - 695). Incubation, larval and pupal periods last for 3 - 9, 9 - 20 (50 - 60 during winter) and 8 - 12 days respectively. A single life-cycle takes 22 - 25 days extending up to 74 days during winter and there may be 8 - 12 generations in a year. *Earias insulana* (Boisduval), is found damaging okra specially in drier regions.

Damage: Eggs are usually laid singly on buds and flowers and occasionally on fruits as well, but in absence of these parts i.e. at early stage of crop growth, the eggs are laid on shoot tips. When the crop is only a few weeks old, the freshly hatched larvae bore into tender shoots and tunnel downwards, these shoots wither, droop down and ultimately the growing points are killed, side shoots may arise giving the plants a bushy appearance. With the formation of buds, flowers and fruits, the caterpillars bore inside these and feed on inner tissues. They move from bud to bud and fruit to fruit thus causing damage to a number of fruiting bodies. The damaged buds and flowers wither and fall down without bearing any fruit whereas the affected fruits become deformed in shape and remain stunted in growth.

Management:

1. Remove debris and all the alternate host plants from field; collect and destroy all the infested shoots and fruits.
2. Grow resistant cultivars like AE 57, PMS 8, Parkins Long green, PKX 9275, Karnual Special
3. Collect and destroy infested shoots, buds, flowers and fruits.
4. Remove the alternate hosts like *Hibiscus cannabinus*, *H. abelmoschus* and *Abutilon indicum* .

5. Release egg parasitoid *T. chilonis* and larval parasitoid *Chelonus blackburnii.*
6. Release first instar larvae of *Chrysoperla carnea* @ 1 lakh/ha.
7. Set up light traps to monitor the moths and their egg laying @ 12/ha
8. Set up pheromone traps @ 5/ha.
9. Seed treatment with Imidacloprid 70 WS @ 10g/kg seed.
10. Spraying of Thiodocarb 75 WP @ 1 g/lit is recommended before fruiting. Spraying of NSKE 5% or Azadirachtin 5% 400 ml or Fenpropathrin 30 EC 250-340 ml or Pyridalyl 10 EC 500-750 ml with 500 L – 700 L water/ha. Emamectin benzoate 5 % SG @3.0 g/10 lit, Spray *B.t* formulation such as dipel @ 2 g / lit.starting from one month after planting at 10 days interval after harvesting the fruits.

2. Red spider mite, *Tetranychus cinnabarinus* (Tetranychidae: Arachnida)

Identification: Eggs are globular in shape, about 0.1 mm in diameter and whitish in colour. Larvae are about 0.2 mm in length and pinkish in colour. Nymphs are greenish-red in colour and about 3 mm in length. Larvae and nymphs look alike in shape but can be easily distinguished as larvae have 3 pairs of legs while nymphs and adults have 4 pairs of legs. There are only two nymphal stages – protonymphal and deutonymphal. Adults are ovate in shape, reddish-brown in colour and 0.4 mm (male) - 0.5 mm (female) in length with four pairs of legs.

Biology: Eggs hatch in 4 to 7 days; larval development takes 3 - 5 days; protonymphal and deutonymphal stages last for 3 - 4 days each. Longevity of adult males and females is 4 - 9 and 9 - 18 days respectively. The females that are active during summer in northern India become active with the onset of monsoon and lay eggs parthenogenetically. These unfertilized eggs give rise to males only but the subsequent generations are sexual.

Damage: Colonies of mites comprising of eggs, nymphs and adults are found feeding on ventral surface of leaves under protective cover of fine silken webs. As a result of their feeding innumerable yellow spots appear on the dorsal surface of leaves and the affected leaves gradually start curling and finally get wrinkled and crumpled. This in turn affects the growth and fruit formation capacity of the plants.

Management:

1. Spray wettable sulphur 2g/l (or) dicofol 3 g/lit .
2. Spray Abamectin 1.9 EC @ 500 ml which is a new acaricide or fenpropathrin 30 EC 250-340 ml in 500 L water/ha.
3. **Cotton jassid :** ***Amrasca devastans*** As observed under "Pests of brinjal".
4. **White fly,*Bemesia tabaci (Aleurodidae, Hemiptera)***

Identification: Nymph is greenish yellow in colour and is present along with pupa on the under surface of the leaves. Adult is minute yellow colour bodied insects with white waxy covering.

Biology: The females mostly lay eggs near the veins on the underside of leaves. Each female can lay about 300 eggs in its lifetime. Egg period is about three to five days during summer and 5 to 33 days in winter. Nymphal period is about 9 to 14 days during summer and 17 to 73 days in winter. Adults emerge from puparia through a T-shaped slit, leaving behind empty pupal cases or exuviae. Adults live from one to three weeks.

Damage: Nymphs and adult remain on the under surface of the leaf and suck the cell sap causing chlorotic spots and yellowing and drying of leaves. It is a vector of *"yellow vein mosaic virus"* (YVMV). The fruits turn white and hard which has no market value.

Management:

1. Remove the wild *Abelmoschus sp* and destroy, since they are the carriers of YVMV.
2. Remove the infested crop and destroy.
3. Use of resistant genotypes, NOL 285 which is absolutely free from YVMV.
4. Use of yellow sticky traps coated with castor oil @ 5 per acre.
5. Spray thiamethoxam 25 WG 100 ml or azadirachtin 5% 400 ml or fenpropathrin 30 EC 250-340 ml with 500 L water/ha.

5. Red cotton bug, *Dysdercus cingulatus* (*Pyrrhocoridae, Hemiptera*)

Identification: Head, pronotum, and corium are uniformly reddish; pronotum with white anterior collar, corium with a rounded central spot not touching inner or outer margins of fore wing ; membrane black with posterior margin white; first visible abdominal sternite with transverse whitish fascia , femora at least apically dark.

Biology: The eggs are laid in clusters of 80-100 in cracks of the soil or dry leaves near the plants. The nymphs hatch out in about 7 days and become adults in 40-85 days. The red coloured nymphs are marked by a row of 3 black spots in the middle of the abdomen and 3 white spots on either margin of it.

Damage: Both nymphs and adults suck the leaf and fruit sap. The plants become weak and stunted, the leaves and fruits may curl up.

Management:

1. Avoid alternate cultivated host crops of the whitefly like brinjal, bhendi, tomato, tobacco and sunflower in the vicinity of cotton crop.

2. Removal and destruction of alternate weed hosts like *Abutilon indicum* and *Solanum nigrum* from the fields and neighboring areas and maintaining field sanitation.
3. Adoption of crop rotation with non-preferred hosts such as sorghum, ragi, maize.
4. Timely sowing with recommended spacing, preferably wider spacing and judicious application of recommended dose of fertilizers, particularly nitrogenous and irrigation management is essential to arrest the excessive vegetative growth and pest build up.
5. Conservation of Reduvid, *Herpactor costalis* which is predaceous on this bugs.
6. Monitoring the activities of the adult white flies by setting up yellow sticky traps at 1 foot height above the plant canopy.
7. Spray any one of the following insecticides when pest population reaches ETL: Imidacloprid 17.8% SL 40 – 50 ml/acre or Buprofezin 25%SC 400 ml/acre or Diafenthiuron 50%WP 240 g/acre or Thiacloprid 21.7%SC 200- 240ml/acre or Flonicamid 50% WG 60 g/acre or Thiamethoxam 25% WG 80 g/acre or Fipronil5% SC 600-800 ml/acre.

CHAPTER 4

PESTS OF CRUCIFEROUS CROPS

Cruciferous vegetables belongs to the family Brassicaceae (also called Cruciferae) with many genera, species, and cultivars being raised for food production such as cauliflower, cabbage, kale, bok choy, broccoli, brussels sprouts, mustard plants etc. Among species of the Brassicaceae family, chromosome numbers vary from n = 4 to 128. India occupies first position in the production of cauliflower and third in cabbage in the world.

Cabbage (*Brassica oleracea* var capitata) is a good sources of protein (1.3%) which contains all essential amino acids, particularly sulphur containing amino acids. Cauliflower (*Brassica oleracea*) is one of the most important vegetable crops of India. The edible part of cauliflower is known as curd. It has high quality of proteins and peculiar in stability of vitamin C after cooking. Cruciferous crops are an excellent source of minerals such as calcium, iron, magnesium, sodium, potassium and phosphorus. They have substantial amounts of β carotene (provitamin A), ascorbic acid, riboflavin, niacin and thiamine. Flavour in cabbage leaves is due to the glycoside sinigrin. Leading states for production of cruciferous crops are Uttarakhand, Uttar Pradesh, Himachal Pradesh, Haryana, Rajasthan, Bihar, Gujarat, Maharashtra, Odisha and Karnataka.

The cultivated area for cauliflower in India spans 0.48 million hectares, resulting in a total production of 9.53 million metric tonnes and a productivity rate of 19.6 metric tonnes per hectare in 2021-22. While the volume of cabbage production in India is estimated to have amounted to about 9.95 million metric tons. Varieties of cabbage like Golden acre, Pusa Drum Head, Pride of India, Copenhagen Market, Pusa Mukta, Pusa Synthetic, Midseason Market, September Early, Early Drum Head, Late Large Drum Head, K-1 and varieties of cauliflower like Pant Gobhi 3, Pant Gobhi 2, Pusa Deepali, Pusa Sharad, Pusa Shubhra, Pusa Himjyoti, Punjab Giant 35, Pusa snowball-1, Pusa snowball K-1, Ooty-1 are released by IARI, New Delhi, and SAU's in India.

Average yield of cauliflower is 12-15t/ha, while hybrids can yield up to 40 t/ha. Whereas, an yield of 25-30 t/ha, 40-60 t/ha can be achieved from early types and late type of cabbage . However, the cruciferous crops production and productivity is limited by insect pests like diamondback moth, head borer, semiloopers, Tobacco caterpillar, cut worms and other sucking pests which are discussed in this chapter along with their IPM practices.

1. Diamond back moth, *Plutella xylostella* (Plutellidae : Lepidoptera)

Identification: The moth is grayish brown with narrow wings having pale white markings anteriorly which from diamond-like white patches dorsally when wings are folded over back at rest. Eggs are yellowish white, oval and flattened. Larva is greenish and short, tapering at both the ends.

Biology: It lays up to 57 eggs singly on the under surface of leaves along the veins. Egg period is 4 – 5 days. The larva is greenish with short thin hairs on the body. The larval period is 13 – 21 days. It pupates in a thin loose mesh of silken cocoon and the pupal period is 7 – 9 days. The period from egg to adult occupies 24 - 35 days.

Damage: The larvae bite holes in leaves and cause serious damage. The larvae cause blisters on the leaves which dry away in course of time. The typical symptoms are withered appearance of the leaves and skeletonisation of the plant.

Management:

1. Mustard sown as trap crop twice i.e. 12 days preceding planting cabbage and again 40 days later controls DBM.
2. Remove and destroy plant remenent, stubbles, after harvest and plough the field.
3. Erection of Pheromone traps @ 12/ha
4. Crop rotation with cucurbits, beans, peas, tomato and melon
5. Larval parasitoid: *Diadegma semiclausm* @ 1,00000/ha (Hills – below 25 –27° C) *Cotesia plutellae* (plains) at 20000/ha release from 20 days after planting.
6. Preservation of *Apanteles ruficrus, A. plutellae, B. excarimata.*
7. Neem seed kernel extract 5%, Spray with Azadirachtin 5% Neem Extract Concentrate @ 5.0 ml/10 lit., Spinosad 2.5 % SC@ 1.2 ml/lit, Trichlorofon 50 % EC @ 1.0 ml/lit., Lufenuron 5.4 % EC@1.2 ml/lit., Larvin 1g /lit, Emamectin benzoate 5 % SG @3.0 g/10 lit, *Bacillus thuringiensis* var *kurstaki* 2g/lit

2. Cabbage butterfly , Pieris brassicae (Linnaeus) (Pieriidae: Lepidoptera)

Identification: Eggs are flask-shaped, about one mm long and yellowish in colour. Full grown caterpillars are 38 - 44 mm long, velvety bluish-green in colour with black dots and yellow dorsal and lateral stripes covered with white hair. Pupae are yellowish-green with black spots and dots. Adult butterflies have snow-white forewings with black distal margins more developed in females than in males; hind wings are also pure white with black apical spots. Wing expanse is 60 - 70 mm. Moths emerging in summer are larger in size than those of winter.

Biology: Eggs are laid in clusters under surface of the leaf. A single female lays only 2 - 3 egg-masses of 50 - 80 eggs each. Incubation, larval and pupal periods are on an average 3.2, 5.6 and 7.3 days during May extending upto 17.6, 40.7 and 28.8 days respectively in January. Generally there are two generations during winter (plains) and 4 - 5 in summer (hilly region).

Damage: On hatching, the young caterpillars feed gregariously on leaves for a couple of days, and then disperse, spreading infestation to the adjacent plants and fields. As a result of their feeding the leaves are skeletonized, sometimes the caterpillars bore into the heads of cabbage and cauliflower.

Management:

1. Pest can be checked by handpicking and mechanical destruction of caterpillars during early stage of attack when the caterpillars feed gregariously.
2. Conservation of parasitoid *Apanteles glomeratus.*
3. Intercropping with fenugreek, lucerne and turnip will reduce Cabbage butterfly.
4. Spraying with commercial formulation of *Pieris rapae* granulosis virus and *Bacillus thuriengensis*, Bypel 1® (PrGV + Bt) (1.5 g/l *w/v*) andaqueous neem kernel extract (ANKE) (50 g/l w/v)
5. At severe infestation, spraying with Newer insecticides like novaluron 10EC 375 ml / ha, spinosad 2.5 SC @ 600 ml / ha, indoxacarb 15.8 SC @ 500 ml /ha and flubendiamide 480 SC @ 75 ml / ha.

3. Cabbage head borer, Hellula undalis (Pyralidae, Lepidoptera)

Identification: Moths are pale greyish-brown, covered with reddish colour. Forewings have grey wavy lines, a apical spot and pale edged dark lunule; hind wings are pale dusky, darker in apical area. Larvae (15-18mm) are pale whitish brown with 4 or 5 pinkish - brown longitudinal stripes.

Biology: The moth lays yellowish shiny eggs on the leaves and they hatch in about 4 days. The larva becomes full grown in about 9 days and pupates in soil. The pupal period lasts for 6 days.

Damage: Caterpillars initially mine the leaves and make it white papery. Latter they feed on leaves and bore into stems, entrance hole is covered with silk and excreta.Finally, they enter into cabbage heads, leaf petiole and stem. In severe case of attack the plants become weak and produce deformed heads.

Management:

1. Collect and destroy mechanically caterpillars in the early stages of attack
2. Preservation of larval parasitoid, Bracon hebator
3. Spray *Bacillus thuringiensis* @ 2g/lit at primordial stage, Cartap hydrochloride @ 500g /ha or malathion 50EC @500ml/ha

4. Cabbage aphid, *Brevicoryne brassicae* (Aphididae: Homoptera)

Identification: Eggs when present are pale yellow with greenish tinge. Nymphs are 1.0 - 1.5 mm long and yellowish-green in colour while adults are 1.8 - 2.0 mm long and darker in colour than nymphs.

Biology: Reproduction is mostly viviparous parthenogentic during summer and mild winter. However, during severe winter sexual reproduction may also occur. Eggs are laid during November –December. These hatch in 20 - 22 weeks. The nymphs mature in about 2 weeks and immediately start producing young ones, without mating. A single female may produce 40 - 45 young ones during her life time. The life cycle is completed in 11 - 45 days and as many as 21 generations have been recorded during a year when provided with favourable conditions.

Damage: Colonies of these insects are often found on tender shoots and as a result of sucking of vital sap from the tissues, the plant remain stunted in growth resulting in poor head formation. In the case of severe infestation plants may completely dry up and die away. When infestation occurs on seedlings, they loose their vigour, get distorted and become unfit for transplanting. The aphids also produce copious quantity of honeydew affected plant parts hindering the photosynthesis and adversely affecting the plant growth.

Management:

1. Ants tend aphids and feed on the honeydew aphids excrete. At the same time, ants protect the aphids from natural enemies. If you see ants crawling up aphid-infested trees or woody plants, put a band of sticky material around the trunk to prevent ants from climbing up.
2. Conservation of predators, *Coccinella septumpunctata, Menochilus sexmaculata,* syrphids and *Chrysoperla carnea.*
3. Weed management.
4. Growing seedlings under protective covers.
5. Silver-colored reflective mulches have been successfully used to reduce transmission of aphid-borne viruses in summer squash, melon, and other susceptible vegetables.
6. Knock off the insects with a strong spray of water.
7. Spray Rogar 30EC @ 2ml/lit (Dimethoate) or Imidachloprid @ 0.25 ml/ lit , Acetamipride @ 1 gm/lit or Thiamethoxam 25% @ WG 0.5 gm/litre of water.

5. Cabbage semilooper, *Trichoplusia ni* (*Noctuidae, Lepidoptera*)

Identification and Biology : Cabbage looper eggs are hemispherical in shape, with the flat side affixed to foliage. They are deposited singly on either the upper or lower surface of the leaf, although clusters of six to seven eggs are not uncommon. The eggs are yellowish white or greenish in color, bear longitudinal ridges, and measure about 0.6 mm in diameter and 0.4 mm in height. Eggs hatch in about two, three, and five days at 32, 27, and 20 C, respectively, but require nearly 10 days at 15 C .Young larvae initially are dusky white, but become pale green as they commence feeding on foliage. They are somewhat hairy initially, but the number of hairs decreases rapidly as larvae mature. Larvae have three pairs of prolegs, and crawl by arching their back to form a loop and then projecting the front section of the body forward. The mature larva is predominantly green, but is usually marked with a distinct white stripe on each side. Dorsally, the larva bears several narrow, faint white stripes clustered into two broad white bands. In some cases the mature larva is entirely green. The adult forewings of the cabbage looper moth are mottled gray-brown in color; the hind wings are light brown at the base, with the distal portions dark brown. The forewing bears silvery white spots centrally: a U-shaped mark and a circle or dots that are often connected.

Damage: Cabbage loopers are leaf feeders, and in the first three instars they confine their feeding to the lower leaf surface, leaving the upper surface intact. The fourth and fifth instars chew large holes, and usually do not feed at the leaf margin. In the case of cabbage, however, they feed not only on the wrapper leaves, but also may bore into the developing head. Larvae consume three times their weight in plant material daily . Feeding sites are marked by large accumulations of sticky, wet faecal material.

Biology: Development time (egg to adult) requires 18 to 25 days when insects are held at 32 to 21 C, respectively. so at least one generation per month could be completed successfully under favourable weather conditions.

Management:

1. Pest can be checked by handpicking and mechanical destruction of caterpillars during early stage of attack when the caterpillars feed gregariously.
2. Conservation of parasitoid *Apanteles glomeratus*.
3. In case of widespread infestation spray with 0.05% dichlorvos or 0.1% Malathion.

6. Painted Bug, *Bagrada cruciferarum* (Pentatomidae: Hemiptera)

Identification: Eggs are oval in shape about one mm long, pale yellow when freshly laid gradually becoming pinkish-orange. Nymphs are beautifully patterned with a mixture of black, white and orange colour, 1.5 - 4.5 mm long depending on their age. Adults are also black and orange colour bugs similar in colour pattern as nymphs – that's why they have earned the common name of painted bugs. Males are 6 - 7 mm long and females 7 - 8 mm.

Biology: Eggs are laid singly or in batches of 2 - 12 on leaves, stems and flower buds. The mating takes place 2 - 6 days after the final nymphal moult and the oviposition commences a week after first mating and may continue intermittently throughout the life span of the female. A single female may lay as many as 230 eggs @ 15 - 20 eggs per day. Eggs and nymphal duration is recorded as 2 - 5 and 18 - 20 days respectively. A single life-cycle is completed in 3 - 4 weeks and adults live for 16 -18 days with 6 - 8 generations in a year.

Damage: Both nymphs and adults suck cell sap from tender plant parts causing yellowing of leaves which gradually dry up and ultimately fall down exposing the plants to secondary invasion of bacteria and fungi. The plants wilt and wither affecting adversely the yield both quantitatively and qualitatively.

Management:

1. Clean cultivation by removing weeds harbouring this pest is imperative for avoiding infestation of these bugs.
2. In case of heavy infestation, spray with triazophos 40 EC @ 1.5 ml/lit and dimethoate 30 EC @ 2.0 ml / lit atleast 7 - 10 days waiting period should be there between treatment and harvest.

7. Mustard Sawfly, *Athalia lugens proxima* Kulg (Tenthredinidae: Hymenoptera)

Identification: Newly hatched grubs are 2 - 3 mm long, smooth, cylindrical and greenish-grey in colour; full grown ones are cylindrical in shape, 16 - 20 mm long and greenish-black in colour. They look and behave like caterpillars but have 8 pairs of prolegs. Adults are 8 - 12 mm long, having dark head and thorax, orange coloured abdomen and translucent smoky wings with black veins. Females have a strong saw-like ovipositor – hence it has been given the popular name sawfly.

Biology: A female lays on an average 35 eggs (20 - 150). Egg period is 6 - 8 days. Grub development takes 21 - 31 days. Pre-pupal and pupal periods last for 3 - 4 and 7 - 10 days respectively. Severe winter is passed in pupal stage and lasts for about 14 weeks. In Northern India there are three generations during cold season. In South India where there is no severe winter, the pest undergoes as many as 10 overlapping generations in a year.

Damage: Eggs are laid singly, mostly during day time and inserted into leaf tissues near the periphery of leaves. On hatching the grubs nibble the margins of tender leaves but later on bite holes in the leaves. Grubs are diurnal in habit and feed generally during early morning and evening hours. With slight disturbance they fall on the soil and feign death.

Management:

1. Hand-picking of grubs which are not active during dawn and dusk .
2. Spraying with spinosad 0.007 per cent, thiodicarb 0.075 per cent.
3. Application of EPN *Steinernema feltiae,* 1.1 x 10^5 infective juveniles/ml, at weekly intervals, significantly reduced the field population of mustard sawfly.
4. Dusting with 5% Malathion is also effective.

8. Tobacco caterpillar, *Spodoptera litura (Noctuidae, Lepidoptera)*

Identification: The egg is round and dirty white. The newly emerged larva is whitish, then turns yellow green an hour after with a pattern of red, yellow, and green lines from the head to the anal region. As the larva grew bigger, the body turns brown with 3 thin yellow lines down the back. The newly emerged larva singled out the leaf veins during feeding while the big larva includes them. The larva hides in the soil at day time and ceaseless when about to molt. . The female is pale brown while the male is darker. Adult moth fore wings have golden and greyish white patches. with wavy white markings . Hind wings white with a brown patch along the margin.

Biology: Females lay 1000-2000 eggs in egg masses of 100- 300 on the lower leaf surface of the host plant. The masses are covered by hair-like scales from the end of the insect's abdomen. Fecundity is adversely affected by high temperature and low humidity (about 960 eggs laid at 30 C and 90% RH and 145 eggs at 35 C and 30% RH). The eggs hatch in about 4 days in warm conditions, or up to 11-12 days in winter. The larvae pass through six instars in 15-23 days at 25-26 C.

Damage: Larva is nocturnal and heavily defoliates the freshly emerged leaves.

Management:

1. Conservation of larval parasitoids viz., *Cosmopolites sp, Eriborous sp, Rogas sp*
2. Collection and destruction of egg masses and larvae.
3. Spray application of Fipronil 80 WG @ 75 g ai/ha, Spinosad 2.5 SC @ 15 g ai/ha. or cartap hydrochloride @ 175 g a.i./ha once at primordial initiation (22 days after planting) and repeated either thrice at 7 days interval or twice at 10 days interval or *Bacillus thuringiensis* var. kurstaki @ 1 kg/ha.

9. Green peach aphid, Myzus persicae

Identification: The eggs measure about 0.6 mm long and 0.3 mm wide, and are elliptical in shape. Eggs initially are yellow or green, but soon turn black. Mortality in the egg stage sometimes is quite high. Nymphs initially are greenish, but soon turn yellowish, greatly resembling viviparous (parthenogenetic, nymph-producing) adults.

Damage: Green peach aphids can attain very high densities on young plant tissue, causing water stress, wilting, and reduced growth rate of the plant. Prolonged aphid infestation can cause appreciable reduction in yield of root crops and foliage crops. However, green peach aphid does not seem to produce the high volume of honeydew observed with some other species of aphids. Blemishes to the plant tissue, usually in the form of yellow spots, may result from aphid feeding. Leaf distortions are not common except on the primary host. In cabbage and cauliflower, poor head formation occurs. In severe cases of attack, crinkling and cupping of the leaves occur. Aphids also excrete copius amount of honey dew in which sooty mould fungus develops resulting in contamination of harvested products.

Management:

1. Ants tend aphids and feed on the honeydew aphids excrete. At the same time, ants protect the aphids from natural enemies. If you see ants crawling up aphid-infested trees or woody plants, put a band of sticky material around the trunk to prevent ants from climbing up.
2. Conservation of predators, *Coccinella septumpunctata, Menochilus sexmaculata,* syrphids and *Chrysoperla carnea.*
3. Weed management.
4. Growing seedlings under protective covers.
5. Silver-colored reflective mulches have been successfully used to reduce transmission of aphid-borne viruses in summer squash, melon, and other susceptible vegetables.
6. Knock off the insects with a strong spray of water.
7. Spray Rogar 30EC @ 2ml/lit (Dimethoate) or Imidachloprid @ 0.25 ml/lit , Acetamipride @ 1 gm/lit or Thiamethoxam 25% @ WG 0.5 gm/litre of water.

CHAPTER 5

PESTS OF CUCURBITS

Cucurbit crops, or plants in the family Cucurbitaceae, consist of melons, gourds, squash and cucumbers with a chromosome number 2n = 7 – 40. Cucurbits are grown extensively in India and other tropical and subtropical regions. *Cucurbitaceae* is a large family of plants, also known as cucurbits, with 130 genera and 800 species. The name of the *Cucurbitaceae* family came from Latin, where the word *corbis* means bottle or basket, cucurbits were used in various ways in the past (e.g., mature vegetables served as containers or even musical instruments). Cucurbits can be consumed raw or pickled, with pickling being a common preservation method using brine, vinegar, and spices.

Cucurbits, a nutritious food family, such as cucumbers, are popular for their low calories, high water content, and fiber, making them ideal for a diet. They are rich in vitamins A, K, C, and potassium, offering various health benefits. Cucurbits have been used in traditional medicine for various purposes. The seeds are used for treating urinary issues, high blood pressure, kidney stones, and prostate disorders, as well as improving skin conditions. Cucurbits contain important chemical compounds like carotenoids, tocopherols, phenols, terpenoids, and saponins. Recently, triterpenoids, specifically cucurbitacins, have gained attention for their biological properties.

They are important cash crops for small landholders and family farmers, and are also popular for private garden cultivation. In 2022, China dominated cucumber and gherkin production worldwide with 81% share, followed by Turkey and India. India cultivates cucurbits on 4.29 million hectares with a productivity of 10.52 tonnes/ha, contributing 5.6% to the country's total vegetable output. Top cucumber-producing states in India include West Bengal, Madhya Pradesh, Haryana, Karnataka, Punjab, Uttar Pradesh, Assam, Andhra Pradesh, Maharashtra, and Odisha, which collectively made up 84.78% of production in the 2021-22 season.

Cucurbit productivity in India is 10.52 tonnes/ha, however insect pests like fruit fly, leaf miners, pumpkin beetles, and sucking pests minimize its production and efficiency by 20–40% every year. The pest identification, biology, nature of damage and their IPM are discussed in detail in this chapter.

1. Fruit flies, *Bactrocera cucurbitae* and *B. ciliatus B. zonata* (Tephritidae : Diptera)

Identification: Adult with a dark spot in each antennal furrow; facial spot round to elongate. Frons - 2-3 pairs frontal setae; 1 pair orbital setae. Thorax: Predominant colour of scutum red-brown. Post pronotal (=humeral) lobe entirely pale (yellow or orange). Notopleuron yellow. Scutum with parallel sided lateral post sutural vittae (yellow / orange stripes) which extend anterior to suture and posteriorly to level of the intra-alar setae. Medial vitta present; not extended anterior to suture. Scutellum yellow, except for narrow basal band. Anepisternal stripe not reaching anterior notopleural seta. Yellow marking on both anatergite and katatergite. Postpronotal lobe (=humerus) without a seta. Notopleuron with anterior seta.Scutum with or without anterior supra-alar setae; with prescutellar acrostichal setae. Scutellum rarely (5%) with basal as well as apical pair of setae.The eggs of *Bactrocera* Size, 0.8 mm long, 0.2 mm wide, with the micropyle protruding slightly at the anterior end. The chorion is reticulate (requires scanning electron microscope examination). White to yellow-white in colour.Third instar larva: Large, length 9.0-11.0 mm; width 1.0-2.0 mm.Head: Stomal sensory organ small, completely surrounded by 6-7 large preoral lobes, some bearing serrated edges similar to oral ridges; oral ridges with 17-23 rows of moderately long, uniform, bluntly rounded teeth; accessory plates numerous, with serrated edges and interlocking with oral ridges; mouthhooks large, heavily sclerotized, each with a small, but well-defined pre apical tooth. Puparium is Barrel-shaped with most larval features unrecognisable, the exception being the anterior and posterior spiracles which are little changed by pupariation. White to yellow-brown in colour. Usually about 60-80% length of larva.

Biology: In nature the population is generally low during dry weather and increases rapidly with adequate rainfall. Pre-oviposition, egg, maggot and pupal periods last for 9 - 21, 1 - 1½, 3 - 9 and 6 - 8 days respectively. During winter the larval and pupal stages are extended up to 3 and 4 weeks, respectively. A single life-cycle is completed in 10 - 18 days but it takes 12 - 13 weeks to complete a single life-cycle in winter. Adult longevity is 2 - 5 months; females live longer than males.. Pupae are 5 - 8 mm long, barrel shaped and brown to ochraceous in colour. The fully fed maggots come out of the fallen fruits and pupate 10 - 15 cm deep in the soil. Where the fruits do not fall, the maggots pupate inside the fruits (which are not common) or come out, drop down and pupate in the soil.

Damage: The female flies puncture the soft and tender fruits with their stout and hard ovipositor and lay eggs below the epidermis @ 4 - 10 eggs per fruit each time. A single female can lay about 200 eggs in her life span of 8 - 10 weeks.

A puncture made by one female is often used by others also for ovipositing and a single fruit may have more than one puncture made by one or more females. On hatching the maggots feed inside on the pulp of fruits and the infested fruits can be identified by the presence of brown resinous juice which oozes out of the punctures made by the flies for oviposition. These punctures also serve as an entry for various bacteria and fungi; as a result, the infested fruits start rotting, get distorted and malformed in shape and fall off from the plants pre-maturely.

Management:

1. To check the damage by these flies, fruits should be harvested before they start ripening.
2. All the fallen and infested fruits should be collected and destroyed to prevent the carry over of the pest.
3. The flies when they congregate and rest on the under surface of large leaves of ribbed gourd may be controlled by spray application of lambdacyhalothrin @ 1ml/lit
4. Spray application of three rounds of profenofos 0.05 % or fenthion 0.1 % at intervals of 15 days commencing from flowering may be useful.
5. Frequent raking of the soil under the vines or ploughing the infested fields after the crop is harvested can help in killing the pupae.
6. Baits prepared with 10% ripe banana, 10% jaggery mixed with 0.1% Malathion used in bait traps was found effective or this bait mixture is to be applied as 200 spot splashes per hectare on the under surface of cucurbit leaves.
7. Use of 0.4 ml methyl eugenol with 1ml of dichlorvos in bait traps was also found effective.
8. Sex pheromone traps can be installed for monitoting and mass trapping of the fruit fly.

2. Snake gourd semilooper, Anadevidia (=Plusia) peponis F. (Noctuidae: Lepidoptera)

Identification: The brownish moth has shiny brown fore wings. The caterpillar is greenish with white longitudinal lines and black tubercles with thin hairs arising on them.

Biology: The female moth lays white spherical eggs on the under surface of leaves. The semilooper caterpillar with humped last abdominal segment measures 35 - 40mm long. It pupates in a thin silken cocoon in leaf fold. The pupa is greenish but turns dark brown before the emergence of the adult moth. Egg, larval and pupal periods last for 4 - 5, 24 - 30 and 7 - 8 days. Its life-history occupies about six weeks.

Damage: The caterpillar cut the leaf partly and rolls it and lives inside the roll. It is a specific pest of snake gourd and larvae defoliate the plants entirely if infestation is serious.

Management:

1. The larvae when found in small numbers may be hand-picked and destroyed.
2. Encourage activity of *Apanteles taragamae, A. plusiae*
3. Spray application of spinosad 45SC @ 0.4 mL/ litre gives significant eduction in semilooper populations.

3. Pumpkin beteles, *Raphidopalpa foveicollis* (red beetle), *Aulacophora cincta* (grey with black border) and *A. lewisii* (=*A. intermedia*) (blue beetle) (Galerucinae : Coleoptera)

Identification: Adult beetles of *R. foevicollis* are 6 – 8 mm long having glistening yellowish red to yellowish brown elytra that are uniformly covered with fine punctures. Adults of *A. cincta* are similar in size and appearance except that the colour of elytra is greyish yellow to brownish grey with distinct palm margin all around. Adults of *A. lewisii* are slightly smaller (5 – 6 mm) with blackish blue elytra.

Biology: The yellowish pink spherical eggs are laid in the soil which turns orange after two days and a beetle may lay 150 - 300 eggs. The egg period is 5 - 8 days. The grubs become full grown in 13 - 25 days and pupate in the soil. The prepupal period is 2 - 5 days. The pupal period ranges from 7 - 17 days. Total life cycle occupies 32 - 65 days. In a year there may be 5 - 8 generations of the insect.

Damage: The beetles bite irregular holes on leaves and also feed on flowers. They prefer young seedlings and tender leaves and the damage at this stage may kill the seedlings. The roots as well as the stem and fruits that come in contact with the soil are damaged by the grubs. Pumpkin is preferred by *R. foveicollis* and sponge gourd by *A. lewisi.* Both the species feed also on snake gourd, pumpkin, cucumber, melon and ribbed gourd.

Management:

1. Cultural practices like clean cultivation and early sowing reduces pest damage.
2. After harvesting deep ploughing of infested field to kill the grub in the soil.
3. Spray application of malathion 50 EC @ 500 ml or dimethoate 30 EC 500 ml or methyl demeton 25 EC@ 500 ml/ ha.
4. Deltamethrin 2.8 EC (1mL/litre) + jaggery (10g/litre) treatment gavesignificantly highest reduction of pumpkin beetles .

5. The integrated module comprising sprayings of DDVP 76% EC @0.75 ml/l at 20 and 30 days after sowing (DAS), Bacillus thuringiensis var. Kurstaki @ 2 g/l at 40 DAS, Imidacloprid 17.8 SL @ 0.30 ml/l at 50 DAS, Lecanicillium lecanii @5 g/l at 60DAS and Azadirachtin 300 ppm @5 ml/l at 70 DAS was most effective in reducing the red pumpkin beetle .

4. Pumpkin leaf caterpillar, *Diaphania indica* (Pyralidae, Lepidoptera)

Identification: The wingspan is about 30 mm. Adults have translucent whitish wings with broad dark brown borders. The body is whitish below, and brown on top of head and thorax as well as the end of the abdomen. There is a tuft of light brown "hairs" on the tip of the abdomen, vestigial in the male but well-developed in the female. It is formed by long scales which are carried in a pocket on each side of the 7th abdominal segment, from where they can be everted to form the tufts. Unfertilized females are often seen sitting around with the tuft fully spread, forming two flower-like clumps of scales, which move slowly to spread their pheromones. Larva is elongate, bright green in colour with two narrow ends, white longitudinal stripes dorsally. Adult is a medium sized moth having white transparent wings with big brown marginal patches. Females have orange coloured hairs at the anal end.

Damage: Larvae scrap the chlorophyll and webs leaves together and feed within. in addition feeds on flowers and occasionally bores into developing fruits

Management:

1. Release of bioagents viz., *Trichogramma chilonis, Dolichogenidea stantoni, Nomuraea rileyi, Beauveria bassiana, Metarhiziumanisopliae and Bacillus thuringiensis subspecies kurstaki)* for the management melon borer, *Diaphania indica.*
2. Spray application of Emamectin benzoate 5% SG @ 0.40 gm/lit.,Malathion 50 EC @ 1 ml/lit or dimethoate 30 EC @ 2 ml/lit.

5. Coccinia gall fly, *Neolasioptera cephalandrae* (Cecidomydae: Diptera)

Identification: Adults are small dark brown coloured fly.

Damage: Gall formation due to attack of the larvae of *Neolasioptera cephalandrae* is common on the stem of *Coccinia indica*. Isolated patches of secondary meristematic centres develop throughout the proliferated tissue and they often differentiate into vascular tissues (either xylem or phloem or both). The larvae bore through the stem and live in lysigenous cavities in the ground tissue. Proliferation of cells of the secondary meristematic centres results in gall formation and the normal vascular pattern is disturbed and vascular bundles crushed. Development of mechanical tissues in the affected region is suppressed. Additional cambial layers with no definite pattern of orientation also develop in the ground tissue and these produce patches of xylem tissue. Larval cavities are irregular and are surrounded by a nutritive tissue rich in protein and polysaccharides.

Management:

1. Clipping of the galls and destroy.
2. Spraying of Profenofos 50 EC @ 2 ml/l or Acephate 75 SP @ 2 gm/ lit or Spirotetramat 0.5 ml / lit or Fipronil 0.3% 1ml /lit SC or Carbosulfan 25% EC 2.5 ml / lit or Chlorantraniliprole 0.4 ml / lit 18.5 % w/w + Neem oil @ 3 ml / lit
3. Fipronil 0.3 GR @ 6-8 kg/ acre.

6. Serpentine leaf miner, Liriomyza trifolii (Agromyzidae, Diptera)

Identification: Eggs are small, creamy white in colour and oval shaped. The maggots are apodous, bright yellow to yellow green in color, measuring 1/6 inch in length. Pupa is distinctly segmented and yellowish brown in colour. Adult fly is yellowish black coloured fly measuring about 1/12 inch of length.

Biology: Eggs are laid in upper or lower surface of the leaf. Eggs hatch in 2-4 days. Larva feeds between the epidermal layers. Larval stage is completed in 6-7 days. Larva on maturity, makes a longitudinal slit in the leaf, comes out and pupates in the leaves or in the soil.

Damage: Larva mines and feeds on the chlorophyll in between the epidermal layers of the leaves. Characteristic symptom of serpentine mining is observed. Severely affected leaves dry and drop. Large population of the pest destroys leaves and affects the growth of plants.

Management:

1. Collect and destroy mined leaves .
2. Spray application of NSKE 5% or neem formulations or Acephate @ 1ml/ lit. or profenophos @ 2 ml/lit

7. Green peach aphid, Myzus persicae.

Identification: Nymphs and adults are yellow in colour. In adults, both winged and wingless forms are present.

Damage: Nymphs and adults are found in colonies of hundred in tender portions of the plant. They suck the sap from tissues leading to curling and crinkling of leaves. In case of severe attack, the leaves curl up, fade gradually and dry up. Sooty mould develops due to the excretion of honey dew giving the plants a dark appearance.

Management:

1. Conservation of predators, *Coccinella septumpunctata, Menochilus sexmaculata, syrphids* and *Chrysoperla carnea.*
2. Spray 300 ml of Metasystox 25EC (Methyl demeton) or 200 ml of Rogar 30EC (Dimethoate) or 75 ml of Dimecran 85SL (phosphamidon) in 100 liters of water.

CHAPTER 6

PESTS OF CHILLIES

Chilli (*Capsicum annum L*), a vegetable and spice crop grown throughout the world, belongs to Solanaceae family originated from Mexico and Guatemala (South America) with 2n = 24 chromosomes (diploid). It is grown worldwide for its commercial, dietary, and therapeutic benefits. India is the largest producer with 1.98 million tonnes and contributes 43% of world chilli production, followed by China, Ethiopia, Thailand, Pakistan and Bangladesh.

Chilli is described in various vernacular names such as Lal mirch (Hindi), Mirapa kaya (Telugu) Lanka, Lankamorich (Bengali), Marcha (Gujarati), Mensina kai (Kannada), Mulaku (Malayalam), Mirchi (Marathi), Lanka (Oriya), Lalmirch (Punjabi), Milagay (Tamil), and Lalmirch (Urdu).

Chilli is rich in vitamins (A, B, C), minerals, and antioxidants like capsaicinoids and carotenoids, sugar, protein, and beta carotene. These natural compounds in chillies make them popular globally due to their health-promoting benefits include pain relief, anti-inflammatory, antimicrobial, antioxidant, anti-ulcer, anti-cancer, memory enhancement, blood sugar and pressure reduction, cholesterol lowering, anticoagulant effects, anti-worm properties and wound healing abilities.

India is the world's greatest producer, consumer, and exporter of chilli producing 7.32 lakh hectares with an annual output of 19.88 lakh tonnes in 2020-21. Major chilli producing states of India are Andhra Pradesh, Telangana, Madhya Pradesh, Karnataka and West Bengal. Chilli is the most prominent crop raised all over India, and various institutions have released a few vital varieties such as Pusa Jwala, Pusa sadabahar, Arka Gagan, Arka Tanvi, Arka Saanvi, Arka Yashasvi (H 8), Arka Tejasvi, Kashi Tej (CCH-4) F1 hybrid, Kashi ratna CCH-12 F1 hybrid, Kashi abha (VR-339), Kashi Sindhuri. Varieties that are released under Dr. YSRHU, Andhra Pradesh *i.e.,* Dr. YSRHU-Kranti, Dr. YSRHU Chaitra, Dr. YSRHU-Tanvee, and Dr. YSRHU-Siri.

India produces between 50 and 60 tonnes per hectare, with a high yield of red and green chillies per acre. Nevertheless, chilli production and productivity is reduced by insect pests like gram pod borer, tobacco caterpillar, thrips and other sucking pests which are discussed in this chapter along with their management practices.

1. Chilli thrips, Scirtothrips dorsalis (Thripidae, Thysanoptera)

Identification: Chilli thrips are pale colored and the lengths of their first and second instar larvae and the pupae are 0.37-0.39, 0.68-0.71 and 0.78-0.80 mm, respectively. Adults are about 1.2 mm long with dark fringed wings and dark spots forming incomplete stripes which appear dorsally on the abdomen .There are numerous microtrichia and dark transverse antecostal ridges on the abdominal tergites as well as sternites.

Biology: The life cycle stages of *Scirtothrips dorsalis* include egg, first and second instar larvae, prepupa, pupa and adult. Gravid females insert the eggs inside plant tissues above the soil surface. The eggs are microscopic (0.075 mm long and 0.070 mm wide), kidney-shaped and creamy white in color. The eggs hatch between two to seven days, depending upon temperature. Larvae and adults tend to gather near the mid-vein or borders of the host leaf.The two larval stages are completed in eight to ten days and the pupal stage lasts for 2.6-3.3 days. The life span of chilli thrips is influenced by the host plant species. Unlike other thrips, pupae of chilli thrips are generally found on leaves, leaf litter or on the axils of leaves, in curled leaves or under the calyces of flowers and fruits.Reproduction in thrips is generally sexual, parthenogenesis is also present.

Damage: Thrips possesses lacerating and sucking mouthparts and cause damage by extracting the contents of individual epidermal cells leading to necrosis of tissue. This changes the tissue color from silvery to brown or black. Chilli thrips create damaging feeding scars, distortions of leaves, and discolorations of buds, flowers and young fruits by feeding on the meristems of the host plant's terminals and on other tender parts above the soil surface. *Scirtothrips dorsalis* has not been reported feeding on mature host tissues. According to Sanap and Nawale (1987), adult and nymphs of *Scirtothrips dorsalis* suck the cell sap of leaves, causing rolling of the leaf upward and leaf size reduction. For example, a heavy infestation of *Scirtothrips dorsalis* in pepper plants changes the appearance of the plant to what is called "chilli leaf curl." Appearance of discolored or disfigured plant parts suggests the presence of *Scirtothrips dorsalis*.A severe infestation of chilli thrips makes the tender leaves and buds brittle, resulting in complete defoliation and total crop loss. Infested fruits develop corky tissues.

The infested leaves curling upward, crumbling and shedding infested buds turning brittle with petiole becoming brown and dropping down, affected fruits showing light brown scars.

Management:

1. Inter crop with *agathi Sesbania grandiflora* to provide shade which regulate the thrips population
2. Do not grow chilli after sorghum – more susceptible to thrips
3. Do not follow chilli and onion mixed crop – both the crops attacked by thrips
4. Sprinkle water over the seedlings to check the multiplication of thrips
5. Treat seeds with imidacloprid 70% WS @ 12 g /kg of seed
6. Apply carbofuran 3% G @ 33 kg /ha or phorate 10 % G @ 10 kg/ha or spray any one of the following insecticide
7. Imidacloprid 17.8 % SL @3.0 ml/10 lit, Dimethoate 30 % EC @1.0 ml/ lit., Emamectin benzoate 5 % SG@ 4 g/10 lit, Fipronil 5 % SC@1.5 ml/lit., Spinosad 45 % SC @3.2 ml/10 lit, Thiacloprid 21.7 % SC @6.0 ml/10 lit. Ethion 50 % EC @2.0 ml/lit., Oxydemeton –Methyl 25 % EC @1.0 ml/lit, Phosalone 35 % EC@2.0 ml/lit.
8. Thrips, *Franklinothrips vespiformis* and *Erythrothrips asiaticus* are predaceous on the insect.

2. Fruit borers, Spodoptera litura, Helicoverpa armigera, Utethesia pulchella (Noctuidae, Lepidoptera)

Damage:

1. Irregular hole is seen in the fruit due to the feeding of *S. litura.*
2. Round neat hole is seen on the fruit due to the feeding of *H. armigera.*
3. Faded pericarp in typical fashion with seeds intact due to the infestation of *U. pulchella.*

Management:

1. Remove the affected plants.
2. Spray application Malathion 50 EC/ Chlorpyriphos 20 EC @ 2 ml/lit. or thiodicarb 200g/ ac.
3. For Spodoptera, spray application of poison baits 5 kg of rice bran + 500 g Carbaryl/ 500 ml Chlorpyriphos + 500 g jaggery + sufficient amount of water is made into small balls and is applied in the evening.

3. Chilli mites, *Polyphagotarsonemus latus* (*Tarsanomidae, Acari*)

Identification and Biology: Adult broad mites are elliptically shaped, but slightly wider at the front than the rear. Females are about 1.5 mm long and males are slightly shorter and more broad. Live specimens are light, translucent yellowish green. A pale white stripe runs longitudinally down the back of the female. Dead specimens are yellowish brown. They have 4 pairs of whitish legs, but the fourth pair of the female adult is greatly reduced. Females live for about 10 days and lay an average of 2 to 5 eggs per day (20 - 50 eggs per female).Without fertilization, females produce eggs which result in only male progeny .

Broad mite eggs are oval in shape and slightly flattened .The exposed translucent surface is covered with five or six rows of white bumps called tubercles. Eggs are about 3/100 inch (0.7 mm) long and can be discerned with a 14X handlens. They are usually laid singly on the undersides of new growth leaves .On fruit eggs are laid on the protected surface or in the depressions of the fruit .Eggs usually hatch in 2 to 3 days.

Larvae are very small, pear-shaped and have three pairs of legs. Just after hatching the larvae are translucent, but females become yellowish green or dark green in color and males yellowish brown .The larvae feed for 1 to 3 days before going into the resting pupal stage.

The pupal stage of this mite is a resting period in which there is no feeding. Sexes are similar in appearance, except for the fourth pair of legs. On males the fourth pair of legs are enlarged; on females, the fourth pair of legs are reduced and whip-like . The pupal lasts 2 to 3 days.

Damage: Mites feed by piercing plant cells and sucking up the sap that oozes from the wound.Reduction in photosynthesis and instability of water balance are some the damaging effects to plants. Feeding damage also causes terminal leaves and flower buds to become cupped and distorted. As a result of feeding injury, corky brown areas appear between the main veins on the underside on the leaf. Young foliage sometimes becomes rust colored and nearly always is deformed. Blooms abort, and the plant growth is stunted. Damaged leaves often turn dark green. *P. latus:* is a vector of leaf curl [Murda disease].

Management:

1. Avoid use of excess nitrogen.
2. Spray application of Dichofol (Kelthane) @ 5 ml/lit, wettable sulphur @ 3 g/lit, micronized sulphur @ 2.5 g/lit.
3. Encourage the activity of predatory mite: *Amblyseius ovalis*
4. Apply phorate 10 % G @ 10 kg/ha or spray any one of the following insecticide

Insecticide	**Dose**
Buprofezin 25 % SC	8.0 ml/10 lit.
Chlorfenapyr 10 % SC	1.5 ml/lit.
Diafenthiuron 50 % WP	8.0 g/10 lit.
Dimethoate 30 % EC	1.0 ml/lit.
Ethion 50 % EC	2.0 ml/lit.

Fenazaquin 10 % EC	2.0 ml/lit.
Fenpyroximate 5 % EC	1.0 ml/lit.
Hexythiazox 5.45 % EC	8.0 ml/10 lit.
Milbemectin 1 % EC	6.5 ml/10 lit.
Oxydemeton –Methyl 25 % EC	2.0 ml/lit.
Phosalone 35 % EC	1.3 ml/lit.
Propargite 57 % EC	2.5 ml/lit.
Quinalphos 25 % EC	1.5 ml/lit.
Spiromesifen 22.9 % SC	ml/10 lit.

4. Green peach aphid: Myzus persicae

Identification and Biology: Nymphs are greenish, but later turn yellowish. Adults are yellowish green in colour. Adults live for 2-3 weeks and produce 8-22 nymphs per day. Both winged and wingless forms breed Parthenogenetically. The Nymphal period lasts for about 7-9 days. It has 12-14 generations per year.

Damage: Appear on the tender shoots, leaves and on the lower surface of the leaves. Suck the sap and reduce the vigour of the plant. Secrete sweet substances which attracts ants and develops sooty mould. The pods that develop black colour due to sooty mould lose quality and fetch low price. The yields are also reduced by aphids directly and more through the spread of virus diseases acting as vectors indirectly.The infested plants turn pale with sicky appearance. The leaves curled and crinkled.

Management:

1. Treat seeds with imidacloprid 70% WS @12 g /kg of seed
2. Acephate 75 sp 1 g/lit or methyl demeton 25 EC 2 ml/lit or phosalone 35 EC 2 ml/lit. Fipronil 5 % SC @1.0 ml/lit., Imidacloprid 17.8 % SL@ 3.5 ml/10 lit .

5. Chilli Blossom midge, *Asphondylia capsici* (Cecidomyiidae: Diptera)

Identification: Fly is dark reddish brown mosquito like midge.

Damage: Fly lays eggs in flower buds. Maggot is tiny pale orange colored and feeds on the floral parts leading to poor development of fruits. The ovary is distorted into gall like structure of varied shape.

Management:

1. Foliar spray with 3 ml/l neem based insecticides, Spirotetramat @ 2ml/lit.
2. Application of acephate 75 SP + DDVP 76 EC @ 250 g+ 250 ml per litre of water .carbosulfan 2 ml/l followed by chlorpyriphos 2.0 ml/l one week later is found effective.

6. Chilli Black thrips , *Thrips parvispinus*

Identification: Body is brown, head and thorax lighter than abdomen; legs and antennal segments III, IV, and V yellow. Forewing brown with pale base. Head wider than long, with small ocellar setae pair III. Antennae with 7 segments, III and IV constricted at apex. Pronotum with 2 long posteroangular setae pairs. Metanotum with median reticulations, weak internal markings, and long median setae. Forewing veins and clavus with setae rows. Abdomen with various setae pairs on tergites and sternites II-VII. Males are smaller and yellowish compared to females. Abdominal tergite II without posteromarginal comb. Pleura tergites lack discal setae.

Biology: Thrips parvispinus reproduces sexually and has multiple life stages: egg, two larval stages, prepupa, pupa, and adult. Females lay an average of 15.33 eggs on young leaves and flowers, with an incubation period of 4-5 days. The duration of each life stage is approximately 4.79 days for egg, 1.36 days for first larval instar, 3.54 days for second larval instar, 1.08 days for prepupa, and 1.96 days for pupa. Females live about 8.55 days, males about 6.00 days, and the entire life cycle takes around 13.68 days.

Damage : Thrips parvispinus, adults and larvae, cause damage by sucking plant sap on flowers and leaf undersides, leading to stunted growth, flower drop, and reduced fruit development. Heavy rainfall leads to a significant infestation in chili crops. Symptoms include punctures, scratches, discoloration, distorted leaf lamina, yellow streaks on leaves, and brown streaks on petals. Severe infestation can dry flowers, impacting fruit set. Thrips feed on plant parts, affecting growth, disrupting pollination by feeding on pollen, primarily near veins on leaves.

Integrated Pest Management (IPM) approaches for T. parvispinus

1. Deep summer ploughing to destroy pupae and residual stages of thrips and other pests.
2. Advance cropping season and avoid staggered planting.
3. Application of well decomposed farm yard manure (FYM) or compost @ 2.5 t/ha, enriched with *Metarhizium anisopliae* or *Pseudomonas fluorescens* @ 2 kg/t along with recommended doses of farm yard manure (25 to 30t/ ha).
4. Soil application of 500 kg of Neem cake and 1.50 -2.00t on of vermi-compost/ha to induce resistance against thrips
5. Seed treatment with Imidachlorpid 70WS @10 g per kg seed.
6. Seedling root dip for 30 minutes with Imidachlorpid 17.8% SL @ 0.5 ml/ L.
7. Follow recommended spacing (60 x 30 cm or 45 x 45 cm) and avoid close spacing, as the high density planting favors the pest incidence and multiplication.

8. Balanced fertilization with enhanced potash application along with nitrogen and phosphorous fertilizers to induce plant resistance against the pest.
9. Mulching with silver coloured polythene sheets of 25–30 micron thickness to reduce pupation of thrips in the soil.
10. Border cropping with 2-3 rows of tall growing crops like sorghum / maize / bajra / fodder grasses etc. sown thickly as a barrier for thrips movement.
11. Intercropping chilli with maize / sorghum and cowpea at10:3:1 as barrier and reservoir crops for natural enemy multiplication, leading to biological control of thrips.
12. Frequent inter cultivation (earthing up/raking of soil) operations to destroy soil inhabiting pupae of thrips
13. Clean cultivation and maintaining weed free bunds are crucial for the management of pests. Uprooting and destruction of weeds such as *Parthenium hysterophorus, Cleomeviscosa, Prosopis sp, Lantana camara,* Calotropis sp, *Tecoma sp., Abutilon sp.,* wild *Solanum sp.,* etc. present in the vicinity of field bunds which act as off season and alternate host for thrips.
14. Crop rotation with crops belonging to the family Poaceae or Gramineae (cereals).
15. Nipping and destruction of severely infested apical shoots at vegetative stage for destruction of thrips residing over apical parts.
16. Mechanical destruction of severely infested plants by uprooting and burying or incineration.
17. Erection of blue or yellow/white sticky traps at65–75 traps/ha at crop canopy height for mass trapping purpose and 20-25 traps/ ha for monitoring purpose.
18. Adopting sprinkler irrigation system instead of flood irrigation, since the jet of water spray from sprinklers disrupts the growth and multiplication of thrips.
19. Spraying of microbial based insecticides like *Beauveria bassiana* or *Lecaniciliumlecanii* at 4.00 g or ml/L (spore load - 1x108 cfu/g or ml), *Pseudomonas fluorescence* – NBAIR-PFDWD at 20g/L or Bacillus albus – NBAIR-BATP @ 20 g/L uniformly covering whole plant.
20. Foliar spray of Entomo-Pathogenic Nematode (EPN), *Steinernema carpocapsae* formulation at10g/L + 1 g wetting agent. Soil application of EPNs, *Steinernema carpocapsaeor, Heterorhabditis indica* at7.50- 12.50 kg/ha. It can be applied as soil drenching after mixing in 500 -750 L of water.

22. Spraying % Neem Seed Kernel Extract (NSKE) or 5% Neem Seed Powder Extract or 0.50% Neem oil (5 ml/L), 0.50% Pongamia oil (5 ml/L), and 5% Vitex negundo extract (50 ml/L). , Spraying of commercial formulation of neem based insecticide (Azadirachtin3000 PPM) @ 2 ml/L.
23. As a final resort the following label claim insecticides can be sprayed judiciously against black thrips .

Insecticide	**Dosage per ha**	**Waiting period (in days)**
Cyantraniliprole 10.26% OD	600 gm in 500 litre	03
Emamectin benzoate 05% SG	200 gm in 500 litre	03
Ethion 50 % EC	1.5-2 litre in 500-1000 litre	05
Fenpropathrin 30 % EC	250-340 ml in 750-1000 litre	07
Fipronil 05 % SC	800-1000 gm in 500 litre	07
Imidacloprid 17.80 % SL	125-250 ml in 500-700 litre	40
Lambda-cyhalothrin05% EC	300 ml in 400-600 litre	5
Spinosad 45 % SC	160 gm in 500 litre	3
Spirotetramat15.31%w/wOD	400 gm in 500 litre	5
Thiamethoxam 30 % FS	This is used as seed dresser	
Tolfenpyrad 15 % EC	1 litre in 500 litre	7
Diafenthiuron 47 % + Bifenthrin 09.40 % w/w SC	625 ml in 500 litre	7
Emamectin Benzoate 01.50 % + Fipronil 03.50 % SC	500-750 gm in 500 ltr.	3
Emamectin benzoate5 % w/w + Lufenuron 40 % w/w WG	60 gm in 500 litre	3
Flubendiamide 19.92 % + Thiacloprid 19.92 % w/w SC	200-250 ml in 500 ltr	5
Fipronil 07 % + Hexythiazox 02 % w/w SC	1 litre in 500 litre	7
Hexythiazox 3.5% + Diafenthiuron 42% WDG	650 gm in 500 litre	7
Indoxacarb 14.5 % + Acetamiprid 7.7 % w/w SC	825-875 ml in 500 ltr.	5
Profenofos 40 % + Fenpyroximate 2.5 % w/w EC	1 litre in 500 litre	7

CHAPTER 7

PESTS OF PERRENIAL AND LEAFY VEGETABLE CROPS

India has a diverse range of leafy greens with untapped potential. They are rich in phytochemicals essential for health, especially for those unable to afford high-quality vegetables. Indigenous leafy greens are cost-effective, resilient, and beneficial for poor communities in remote areas. Popular leafy vegetables in India include Indian spinach, amaranthus, fenugreek, drumstick, and curry leaf, along with lesser-known seasonal varieties.

Leafy vegetables, known as 'Saag' in India, are tied to social and cultural heritage. For instance, purslane is used during Jivitputrica festival in Bihar, while 14 leafy vegetables are cooked on 'Bhoot Chaturdashi' in Bengal. Beary Muslims in Karnataka make Basalede kunhi pindi and Bengali cuisine features poi in various dishes. In Maharashtra, greens are popular during festivals like Rishi Panchami, preserving biodiversity through traditional culinary practices.

In Andhra Pradesh, Basella and Yam curry is known as Kanda Bachali Koora, while Gongura, also known as roselle plant, is grown for its sour leaves in India and Fiji. Gongura comes in green and red stemmed varieties, with the red one being sourer. Rich in iron, vitamins, folic acid, and antioxidants, Gongura is essential in Andhra cuisine, especially in the form of Gongura pacchadi. Known as 'mother Andhra', Gongura is a popular ingredient in pickles, homes, hotels, and restaurants across Andhra Pradesh.

Green leafy vegetables are nutrient-dense and offer a variety of health benefits with their high mineral, vitamin, and antioxidant content. They help manage chronic diseases like cancer and heart issues, while also preventing anemia and aiding bone health. Some varieties like amaranth and Purslane are high in protein and essential amino acids. Indian pennywort and Water Pennywort have medicinal properties for memory enhancement, skin issues, and liver support. Incorporating leafy greens into your diet can positively impact your health and overall well-being.

Important varieties of green leafy vegetables released in India - Amaranth (Arka Suguna, Arka Arunima, Arka Samraksha, Arka Varna, Pusa kirti,Co.3), Palak (Arka Anupama, Pusa Jyothi, Pusa Bharathi), Fenugreek (Kasuri Supreme, Kasuri methi, Pusa early bunching, Hisar suvarna). Nevertheless, green leafy vegetables production and productivity is limited by insect pests like moringa Hairy caterpillar , moringa Leaf eating caterpillar , amaranthus caterpillar , amaranthus weevil other pests which are discussed in this chapter along with their management practices.

1. Moringa Hairy *caterpillar,Eupterote mollifera* (*Eupterotidae, Lepidoptera*)

Identification and Biology: Adults are large moths with yellowish wings having faded lines. Moths emerge with inception of monsoon and lay eggs in groups on leaves and tender stems. Egg period is 6 days. Matured larva are brownish and densely hairy. Hairs are annoying to handle. Pupation takes place in soil.

Damage: Larvae feed gregariously by scrapping bark and defoliating foliage. Severe invasion results in absolute defoliation of the tree.

Management:

1. Collect and destroy egg masses and caterpillars.
2. Use light traps to attract and kill adults @ 1/ha.
3. The larva in groups is killed by burning torch.
4. Spray application of Azadirachtin 1.0 % EC (10000 ppm) @2.0 ml/ lit., Chlorantraniliprole 18.5 SC @ 60 ml/ha, Emamectin benzoate 5 % SG @4 g/10 lit.

2. Moringa Leaf eating caterpillar, Noorda bliteallis (Crambidae, Lepidoptera)

Identification: Caterpillar is brown colour with a noticeable mid dorsal stripe and a black head . Adult is tiny moth with dark brown fore wings and white hind wings with brown border.

Damage: Caterpillar webs the leaves and feed on the leaflets further reducing them into papery structures.

Management:

1. Deep Plough the soil around the trees to expose and kill pupae.
2. Spray application of Malathion 50 EC @ 2 ml/lit., Neem oil 3ml/lit @ 1g/ lit., or Fipronil 5 % SC@1.5 ml/lit.

3. Moringa Bud midge,Contarinia moringae (Cecidomyiidae, Diptera)

Identification: Adult fly is a small brownish in colour.

Damage: Adult Female lays eggs in batches on anthers of the bud. Maggot nourish on the internal contents of the flower buds and causes shedding of flower buds.

Management:

Spray application of Malathion 50 EC @ 2 ml/lit., Fipronil 5 % SC@1.5 ml/lit.,Dichlorvos 76 WSC @ 1 ml/lit, Imidacloprid 17.8 % SL @3.0 ml/10 lit, Dimethoate 30 % EC @1.0 ml/lit., Emamectin benzoate 5 % SG@ 4 g/10 lit,

4. Amaranthus caterpillar, Hymenia recurvalis (Pyraustidae, Lepidoptera)

Identification: Wingspan of both sexes 22-24 mm, forewings are 9-11 mm long, deep brown with broad white median band; hind wings deep brown with a broad white median bar. Mature larva is 15 mm long; head light yellowish-brown with many brown spots, body appears green with transparent epidermis.

Damage: Larvae skeletonize leaves of leafy vegetables .Affected portion dry up.

Managemet: Spray application of Dichlorvos 75 WSC @1.5 ml/lit.

5. Amaranthus weevil,Hypolixus truncatulus (Curculionidae, Coleoptera)

Identification: Grub is apodous stout, "C" shaped, white in colour. Adult weevil with conspicuous snout, grey in colour.

Damage: Larvae bore into the stems of amaranthus and girdle the stems. In affected plants, gall like swellings can be observed on the stem which is scarred with longitudinal white markings. Inside the stem, tunneling by stout creamy white larva can be seen with frass. In severe cases, plants show yellowing of leaves and wilting.

Management: Spray application of Spray Azadirachtin 1.0 % EC (10000 ppm) @2.0 ml/ lit, Fipronil 5 % SC@1.5 ml/lit., Spinosad 45 % SC @3.2 ml/10 lit.

CHAPTER 8

MCQ ON IPM AND VEGETABLE CROP PESTS

[JRF/SRF/NET/ARS/SAU EXAMS]

INTEGRATED PEST MANAGEMENT

1. In neem , the insecticidal property is chiefly due to the bioactive principle called [ ]

   a. *Citrinellol* b. *Azadirachtin*

   c. *Garaniol* d. Eugenol

2. Inundative release is an activity associated with [ ]

   a. *Chemical Control* b. *Mechanical*

   c. *Cultural Control* d. *Biological Control*

3. DDT was discovered in [ ]

   a. 1947 b. 1939

   c. 1941 d. 1942

4. Malarial parasite was first discovered by Sir Ronald Ross in [ ]

   a. Paris b. Madras

   c. Calcutta d. Hyderabad

5. Commercial formulation of Bacillus thuringiensis is [ ]

   a. DOOM b. DIPEL

   c. GIPLURE d. VAPONA

6. Chlorpyriphos is a [ ]

   a. Synthetic pyrethroid b. Carbamate

   c. Organophosphate d. Chlorinated hydrocarbon

7. A common antidote for insecticide poisoning is [ ]
   a. Nicotine b. Methoprene
   c. Moban d. Atropine
8 Genetic Engineering Appraisal Committee [GEAC] works under [ ]
   a. Ministry of Environment and Forest
   b. Ministry of Agriculture
   c. Ministry of Commerce
   d. Ministry of Science
9. In India, Bt cotton is released for commercial cultivation by GEAC during the year [ ]
   a. 1998 b. 2000
   c. 2002 d. 2005
10. Directorate of Plant Protection, Quarantine and storage is located at [ ]
    a. Faridabad b. Dehradun
    c. Jodhpur d. Hyderabad
11. Biodiversity act was passed in the year [ ]
    a. 1998 b. 2000
    c. 2002 d. 2005
12. In which year GOI passed Insecticide Act ? [ ]
    a. 1959 b. 1969
    c. 1958 d. 1968
13. Destructive Insect and Pests Act [DIPA] was passed in the year [ ]
    a. 1903 b. 1914
    c. 1929 d. 1968
14. National Bureau of Agriculturally Important Insects [NBAII] is located at [ ]
    a. New Delhi b. Chennai
    c. Bangalore d. Mau
15. Antixenosis term was proposed by which scientist ? [ ]
    a. R.H.Painter b. Kogan & Ortman
    c. M.S.Swaminathan d. E.F.Knipling
16. On contact with moisture, celphos tablets releases which gas ? [ ]
    a. Phosgene gas b. CO_2
    c. Phosphine d. Sulphur di oxide

17. *Which of the following insecticide belong to phenyl pyrazole* group? [ ]
    a. Fipronil b. Indoxacarb
    c. Etofenprox d. Imidachloprid
18. Active principle of BHC is [ ]
    a. Hexa chloro b. Hexa chloro hexane
    c. Hexa hexane d. Hexa chloro pentane
19. *Microsomal enzyme responsible for metabolism of Organi* Phosphate insecticides [ ]
    a. Esterases b. Cytochrome P 450
    c. Glutathion S Transferase d. All the above
20. White grubs can be controlled by [ ]
    a. Ha - NPV b. *Metarhizium anisopliae*
    c. *Trichogramm chilonis* d. *Apanteles obliqua*
21. Which of the following insects is controlled by NPV [ ]
    a. Spodoptera litura b. Helicoverpa armigera
    c. Red hairy *caterpillar* d. *All of the above*
22. First time quarantine started in India at [ ]
    a. New Delhi b. *Mumbai*
    c. *Kolkotta* d. *Chennai*
23. Biological control agent of *Eichornia crassipes* [water hyacinth] is [ ]
    a. Trichogramma japonicum b. *Bacillus thuringiensis*
    c. *Zygogramma bicolorata* d. *Neochetina bruchi*
24. Which of the following is not correct match [ ]
    a. Phorate - Dursban b. *Aldicarb - Temik*
    c. *Indoxacarb - Avaunt* d. *Triazophos - Hostathion*
25. An insecticide which is used as seed treatment against sucking insect pests [ ]
    a. Imidachloprid b. *Endosulphan*
    c. *Deltamethrin* d. *Monocrotophos*
26. *National Institute of Plant Health Management is located at* [ ]
    a. New Delhi b. Mumbai
    c. *Hyderabad* d. Chennai
27. Tobacco is a trap crop in castor cultivation against [ ]
    a. *Spodoptera litura* b. *Helicoverpa armigera*
    c. *Agrotis ipsilon* d. *Aceria cajani*

28. The fungus, *Verticillium lecanni* is well documented entomopathogen of [ ]
    a. Leaf footed bugs
    b. Leaf and tree hoppers
    c. Stem borers
    d. Thrips, White fly and Scale insects
29. The nymphal - adult *parasitoid, Epiricarinia melanoluca* parasitize [ ]
    a. *Nilaparvata lugens* b. *Alerolobus barodensis*
    c. *Amrasca bigutulla* d. *Pyrilla perpusilla*
30. Trichocard is used for the control of [ ]
    a. *Sucking insect pests* b. *Coleopteran borers*
    c. *Lepidopetran borers* d. *Forest insects*
31. The common inhibitory neurotransmitter substance in the insect nervous system is [ ]
    a. *Acetyl choline* b. *Serotonin*
    c. *Gama- aminobutric acid* d. *Dopamine*
32. *A mechanical device pointed iron hook is used for collecting* [ ]
    a. *Coconut black headed caterpillar* b. *Red palm weevil*
    c. *Rhiniceros betle* d. *Lady bird beetles*
33. *The most commonly used chitin synthesis inhibitors for hemipterans* [ ]
    a. Buprofenzin b. Diflubenzuron
    c. Flufenozuron d. Teflubnzuron
34. Marigold is used as a trap crop against [ ]
    a. Spodoptera litura b. *Helicoverpa armigera*
    c. *Red hairy caterpillar* d. *All of the above*
35. Token stimulus theory was given by [ ]
    a. R.H.Painter b. Kogan & Ortman
    c. Kennedy and Booth d. Fraenkel
36. Resting potential of a nerve cell is [ ]
    a. + 70 mv b. - 60 mv
    c. - 70 mv d. + 50 mv
37. Attractant is type of [ ]
    a. Kairamone b. Pheromone
    c. *Allomone* d. *Both a & b*
38. The weed prickly pear, Opuntia dilleni is controlled by [ ]
    a. *Teleonomus scrupulosa* b. *Dactylopius opuntiae*
    c. *Zygogramma bicolorata* d. *Neochtina bruchi*

39. Biological control agent for *Parthenium hysterophorus* is [ ]
   a. *Ophiomyia lantanae* b. *Dactylopius opuntiae*
   c. *Zygogramma bicolorata* d. *Neochtina bruchi*
40. Timely planting of sorghum escapes the damage by [ ]
   a. *Shoot fly* b. *Shoot bug*
   c. *Stem borer* d. *Sorghum midge*
41. Neem seed kernel extract [NSKE] contains [ ]
   a. *Nicotie* b. *triterpinoids*
   c. *Hexanotriterpinoids* d. *Neonicotinoids*
42. The *intracellular crystal proteins of bacillus thurienginsis* are soluble in [ ]
   a. *Neutral medim* b. *Acidic medium*
   c. *Saline medium* d. *Alkaline medium*
43. *The head quarters of Locust warning organization is located at* [ ]
   a. Bikaner b. Jaisalmer
   c. *Jodhpur* d. *Jaipur*
44. Methyl eugenol is used as attractant against [ ]
   a. *Oriental fruit fly* b. *House fly*
   c. *Melon fruit fly* d. *Carrot fly*
45. Mode of action of insecticide Fipronil is [ ]
   a. Blockage of Chloride ion conduction
   b. Stimulation of Chloride ion conduction
   c. *Disruption of ATP Production*
   d. *Inhibition of sodium ion entry*
46. Antagonistic crop against nematode pest is [ ]
   a. *Marigold* b. *Castor*
   c. *Coriander* d. *Maize*
47. *Host plant resistance where many minor genes are involved and termed quantitative is* [ ]
   a. *Immunity* b. *Vertical resistance*
   c. *Induced Resistance* d. *Horizontal resistance*
48. Compounds that antagonizes the insect Juvenile hormone activity is [ ]
   a. Plumbagin b. *Ecdysone*
   c. *Juvabione* d. *Precocene*
49. *Bacillus thuringiensis is a* [ ]
   a. *Gram negative, aerobic bacteria*
   b. *Gram negative, anaerobic bacteria*

c. *Gram positive, aerobic bacteria*
d. *Gram positive, anaerobic bacteria*

50. Second generation neo-nicotinoid insecticide, acetamiprid is [ ]
 a. *Ach esterase agonist* b. *Dopamine inhibitor*
 c. *Ach receptor agonist* d. *Octapamine inhibitor*

51. Anti –juvenile hormone, isolated from plant *Ageratum houstonianum* is a [ ]
 a. *Protocene* b. *Trichocecne*
 c. *Dicosene* d. *Precocene*

52. Chemical released by an organism which is advantageous to the recipeint [ ]
 a. *Allomone* b. *Kairomone*
 c. *Apneumones* d. *Synamones*

53. *Need based application of insecticides is based on* [ ]
 a. *Population density of the insect* b. *Economic injury level*
 c. *Extent of damage* d. *Economic threshold level*

54. Protection and maintenance of existing natural enemies under biological control is termed as [ ]
 a. *Augmentation* b. *Conservation*
 c. *Natural Control* d. *Both a & b*

55. Following is reported to be resistant to practically to all group of conventional insecticides [ ]
 a. Spodoptera litura b. *Helicoverpa armigera*
 c. *Bemisia tabaci* d. *Plutella xylostella*

56. Mode of action of Avermectins [ ]
 a. *Blocking the Na + gates*
 b. *Inhibition of Ach esterase*
 c. *Stimulation of octopomine receptors*
 d. *Agonist at GABA binding sites*

57. *Indoxacarb belongs to the group of* [ ]
 a. *Organiphosphates* b. *Synthetic pyrethoids*
 c. *Neonicotinoids* d. *Oxadiazine*

58. *Egg larval -Parasitoids* [ ]
 a. *Cotesia plutella* b. *Diadegma semiclasum*
 c. *Chelonus blackburni* d. *Ooencyrtus sp.*

59. What is the mechanism of resistance to synthetic pyrethroids [ ]
   a. *Reduced penetration* b. *Decreased nerve sensitivity*
   c. *Enhanced metabolism* d. *Storage and excretion*

60. Immature of the Neuroptera would be classified as [ ]
   a. Predators b. Parasites
   c. *Herbivores* d. *Scavengers*

61. Insecticide responsible for death of over 100 persons in kerala in 1958 [ ]
   a. Carbaryl b. Endosulfan
   c. Synthetic Pyrethroids d. Methyl Parathion

62. The Bt crystal toxin are classified based on the [ ]
   a. Lipid sequence b. Sterol requence
   c. Amino acid sequence d. N - Sequence

63. Chitin synthesis inhibitor of plant origin [ ]
   a. Dimilin b. Plumbagin
   c. Teflubenzuron d. Buprofezin

64. Single dose anticoagulant rodenticide is [ ]
   a. Warfarin b. Zinc phosphide
   c. Bromodiolone d. All the above

65. Zero ETL is used for the management of [ ]
   a. *Stem borers* b. *Vectors*
   c. *Pests* d. *All the above*

66. The international centre for insect physiology and ecology is located at [ ]
   a. USA b. Germany
   c. India d. Kenya

67. Endogram is the trade name of [ ]
   a. Bt formulation b. *Trichogramma chilonis*
   c. Ha - NPV d. Sl – NPV

68. Number of Aluminum Phosphide tablets required for 100 cubic meter space is [ ]
   a. 120 b. 160
   c. 140 d. 180

69. Which of the following insecticide shows negative temperature correlation [ ]
   a. *Malathion* b. *Parathion*
   c. *DDT* d. *All the above*

70. *Which of the following is used as biocontrol agent against management of Parthenium* [ ]
    a. *Zygogramma bicolorata* b. *Apantles partheni*
    c. *Trichogramma chilonis* d. *All the above*
71. Which of the following is used as an antidote for organochlorines [ ]
    a. *Phenobarbitol* b. *Atropine*
    c. *Phytostigmine* d. *Charcoal*
72. The chemical present in mosquito coils is [ ]
    a. *Allethrin* b. *Endosulfan*
    c. *Cypermethrin* d. *Deltamethrin*
73. Insecticide Act came into force during [ ]
    a. 1971 b. 1981
    c. 1968 d. 1969
74. *Which of the following is synthetic analogue of neries toxin* [ ]
    a. *Hydropene* b. *Methoprene*
    c. *Naled* d. *Cartap*
75. Resurgence of whitefly is largely associated with [ ]
    a. *E xcessive use of DDT*
    b. *Indiscriminate use of synthetic pyrethroids*
    c. *E xcessive use of Endosulfan*
    d. *E xcessive use of Monocrtophos*
76. Which of the following is an emulsifier in insecticidal formulation [ ]
    a. *Teepol* b. *Triton X - 100*
    c. *Polyethethylene Glycol* d. PBO
77. The dose of *HaNPV* required for the control of *Helicoverpa armigera* [ ]
    a. *250 LE* b. *500 LE*
    c. *750 LE* d. *1000 LE*
78. The relation between the following two parameters is a straight line [ ]
    a. Corrected mortality and dose
    b. Probit % mortality and square root of dose
    c. Dose and Probit % mortality
    d. Log dose and Probit % mortality
79. Inclusion bodies are characteristic feature of [ ]
    a. *Bacillus thuringiensis* b. *Beauveria bassiana*
    c. *Baculovirus* d. *Metarhizium anisopliae*

80. *Antibiosis is* [ ]
   a. *Bad Host*
   b. *Expelling guests*
   c. *Adverse effect of host plant on insect*
   d. *All the above*

81. Particle size of aerosols in microns is [ ]
   a. 1- 50 b. 0.1 - 1
   c. 51 - 100 d. 0.01 – 0.1

82. Which of the following insecticide is ecdysone agonist that effect moulting [ ]
   a. *Plumbagin* b. *Juvenoid*
   c. *Methoxyfenozide* d. *Spinosad*

83. Avermectin B1 a is effective against [ ]
   a. *Insects* b. *Mites*
   c. *Nematodes* d. *All the above*

84. First synthesis inhibitor developed as insecticide is [ ]
   a. *Diflubenzuron* b. *Triflumuron*
   c. *Flufenoxuron* d. *Hexaflumuron*

85. Which of the following insecticide is banned for use in vegetable crops [ ]
   a. *Cypermethrin* b. *Methyl demeton*
   c. *Profenophos* d. *Monocrotophos*

86. Acceptable daily intake of pesticide is expressed in [ ]
   a. *Mg/kg body weight /day* b. *Mg/person/ day*
   c. *ppm* d. *None of these*

87. Which transgenic crop occupies largest area under cultivation in the world [ ]
   a. *Cotton* b. *Soybean*
   c. *Corn* d. *Canola*

88. The transgenic crop with protease inhibitor gene inhibit [ ]
   a. *Trypsin* b. Chymotrypsin
   c. *Pepsin* d. *Cathepsin*

89. The *parasporal crystalline toxin in the Bacillus thruringiensis is* [ ]
   a. Protien b. *Carbohydrate*
   c. *Protease inhibitor* d. *Amylase inhibitor*

90 *Bacillus thuringiensis was first isolated by Berliner from* [ ]

a. *Mediterranean flour moth* b. *American bollworm*

c. *European corn worm* d. *Bonbyx mori*

91. Dose killing 50 per cent of treated organism is termed as [ ]

a. DT_{50} b. LC_{52}

b. IP_{50} d. KD_{50}

92. The book "Silent Spring" was written by Rachel Carson in the year [ ]

a. *1952* b. *1972*

c. *1962* d. *1982*

93. Indian Standards for Commercialization of HaNPV & SLNPV is [ ]

a. 1 X 10^9 POB/ML b. 4 X 10^9 POB/ML

c. 1 X 10^4 POB/ML d. 5 X 10^9 POB/ML

94. Indian Standards for Commercialization of Entomopathogenic bacteria [ ]

a. Minimum of 1 X 10 8 CFU/ml or gm

b. Minimum of 2 X 108 CFU/ml or gm

c. Minimum of 2 X 10 9 CFU/ml or gm

d. None of the above

95. Indian Standards for Commercialization of Entomopathogenic fungi [ ]

a. Minimum of 1 X 10 5 CFU/ml or gm

b. Minimum of 1 X 10 8 CFU/ml or gm

c. Minimum of 1 X 10 7CFU/ml or gm

d. None of the above

96. Spinosad, a mixture of spinosyn A and spinosyn D, secondary metabolites of [ ]

a. *Saccharopolyspora spinosa* b. *Bacillus thuringiensis*

c. *Azadiracta indica* d. *Beaveria bassiana*

97. Milbemectin [mixture of milbemycin A3 and milbemycin A4, natural products isolated from the fermentations of *Streptomyces hygroscopicus* subsp. *aureolacrimosus*.] is [ ]

a. *Insecticide* b. *acaricide*

c. *Fungicide* d. *Herbicide*

98. Bagging / wrapping of pomegranate and mango fruits in paper bags avoids the infestation of pomegranate butterfly *Virachola isocrates* and mango fruit fly *Bactrocera dorsalis* is a [ ]

a. *Cultural method* b. *Physical Method*

c. *Mechanical Method* d. *None of the above*

99. Use of an alkathene band around the tree trunks of mango to check the migration of [ ]

a. mealybugs and red ants.
b. *Only Mealy bugs*
c. *Only Red ants*
d. *None of the above*

100. Light trap arrangement for mass trapping and destruction of insect pest is a [ ]

a. *Cultural method*
b. *Physical Method*
c. *Mechanical Method*
d. *None of the above*

PESTS OF VEGETABLE CROPS

1. Withered terminal shoots , shedding of flower buds, bore holes on brinjal fruits plugged with excreta is characteristic feature of [ ]

   a. *Leucinodes arbonalis* b. *Euzophera perticella*

   c. *Urentius histricellus* d. *Amrasca devastans*

2. Pusa purple long is a resistant variety of brinjal against [ ]

   a. *Amrasca devastans* b. *Bemisia tabacii*

   c. *Earias vitella* d. *Leucinodes arbonalis*

3. A lady bird beetle with 28 black spots on elytra [ ]

   a. *Batocera rufomaculata*

   b. *Cylas formicarius*

   c. *Henospilachina vigintioctopunctata*

   d. *Coccinella septumpunctata*

4. Ladder like window and skeletonization of brinjal leaf is due to infestation of [ ]

   a. *Earias vitella*

   b. *Henospilachina vigintioctopunctata*

   c. *Earias insulana*

   d. *Leucinodes arbonalis*

5. Yellow sticky traps are installed in brinjal field against [ ]

   a. *Euzophera perticella*

   b. *Leucinodes arbonalis*

   c. *Amrasca devastans*

   d. *Henospilachina vigintioctopunctata*

6. Vector of " little leaf of brinjal" is [ ]

   a. *Amrasca devastans* b. *Hishimonas phycitis*

   c. *Thrips tabacii* d. *Bemisia tabacii*

7. A serious pest of Bhendi where moth is buff coloured forewing with a green wedge [ ]

   a. *Earias vitella* b. *Helicoverpa armigera*

   c. *Earias insulana* d. *Spodoptera litura*

8. Adult moth with completely green fore wing is pest of bhendi [ ]

   a. *Earias vitella* b. *Diacrisia obliqua*

   c. *Earias insulana* d. *Spodoptera litura*

9. Red spider mite infesting okra is [ ]
   a. *Aceria cajani* b. *Polyphagotarsonemus* latus
   c. *Aceria guerreronnis* d. *Tetranichus cinnabarinus*
10. Bhendi Yellow vien mosaic virus is transmitted by [ ]
   a. *Amrasca devastans* b. *Hishimonas phycitis*
   c. *Thrips tabacii* d. *Bemisia tabacii*
11. Scientific name of diamond back moth is [ ]
   a. *Agrotis ipsilon* b. *Helicoverpa armigera*
   c. *Plutella xylostella* d. *Pieris brassicae*
12. Greyish brown moth with diamond – like white patches dorsally when wings are folded over back at rest [ ]
   a. *Spodoptera litura* b. *Trichoplusia ni*
   c. *Plutella xylostella* d. *Pieris brassicae*
13. Scientific name of cabbage butterfly is [ ]
   a. *Helicoverpa armigera* b. *Pieris brassicae*
   c. *Plutella xylostella* d. *Spodoptera litura*
14. Adult butterflies with snow white forewings with black distal margins [ ]
   a. *Bactrocera dorsalis* b. *Pieris brassicae*
   c. *Myzus persicae* d. *Agrotis ipsilon*
15. Mustard is sown as trap crop in cabbage for the control of [ ]
   a. *Spodoptera litura* b. *Trichoplusia ni*
   c. *Plutella xylostella* d. *Pieris brassicae*
16. Common name of *Hellula undalis* is [ ]
   a. *Cabbage butterfly* b. *Cabbage semilooper*
   c. Cabbage head borer d. Cabbage aphid
17. *Deformed heads of cabbage are due to damage by* [ ]
   a. *Cabbage butterfly* b. *Cabbage semilooper*
   c. *Cabbage head borer* d. *Cabbage aphid*
18. *Athalia lugens proxima is pest on* [ ]
   a. *Rice* b. *Brinjal*
   c. *Cotton* d. *Cabbage*
19. *Athalia lugens proxima has _______________ number of prolegs* [ ]
   a. 4 b. 6
   c. 8 d. 2

20. *Adults are reddish brown flies with yellow markings on thorax and brown bands on the wings* [ ]

    a. *Musca domestica* b. *Bactrocera cucurbitae*

    c. *Liriomyza trifolii* d. *Bemisia tabacii*

21. Methyl eugenol is used as attractant against [ ]

    a. *Bactrocera cucurbitae* b. *Pieris brassicae*

    c. *Plutella xylostella* d. *Bemisia tabacii*

22. Frequent raking up of the soil under the cucurbit vines controls [ ]

    a. Bag worms b. Fruit flies

    c. White flies d. Leaf miners

23. Early sowing of pumpkin will control [ ]

    a. Bag worms b. Fruit flies

    c. Pumpkin beetles d. Leaf miners

24. *Scientific name of pumpkin leaf caterpillar* [ ]

    a. *Plusia peponis* b. *Diaphania indica*

    c. *Bactrocera cucurbitae* d. *Plutella xylostella*

25. The maggots bore inside the tender stem of *coccinia indica* forming galls in distal shoots [ ]

    a. Diaphania indica b. *Alacophora cincta*

    c. *Neolasioptera cephalandrae* d. *Plusia peponis*

26. *Phytomyza horticola* is [ ]

    a. Pea pod borer b. Pea stem fly

    c. Pea leaf miner d. Pea blue butterfly

27. Infested field looks like a cattle grazed field due to attack of [ ]

    a. *Etiella zinkenella* b. *Agrotis segetum*

    c. *Phytomyza horticola* d. *Plusia peponis*

28. *Lampides boeticus is a* [ ]

    a. Pea pod borer b. Pea stem fly

    c. Pea leaf miner d. Pea blue butterfly

29. The larva mines into the developing Potato tubers and makes galleries [ ]

    a. *Spilosoma obliqua* b. *Pthorimaea operculella*

    c. *Cylas formicarius* d. *Agrotis segetum*

30. Adult weevil is small, ant like bluish black in colour with reddish brown prothorax and long snout [ ]
    a. *Oryctes rhinoceros*
    b. *Cylas formicarius*
    c. *Rhyncophorous ferrugenius*
    d. *Myllocerus discolour*
31. *Scientific name of Amaranthus web worm* [ ]
    a. *Hymenia recurvalis*
    b. *Trichoplusia ni*
    c. *Phytomyza horticola*
    d. *Agrotis segetum*
32. *Eupterote mollifera is* [ ]
    a. Moringa bud midge
    b. Moringa pod fly
    c. Moringa hairy caterpillar
    d. Moringa leaf eating caterpillar
33. *Shedding of moringa flower buds is due to infestations of* [ ]
    a. Moringa bud midge
    b. Moringa pod fly
    c. Moringa hairy caterpillar
    d. Moringa leaf eating caterpillar
34. Drying , splitting of moringa fruits from tips and gummy exudates oozing from fruits is due to infestations of [ ]
    a. *Noorda blitealis*
    b. *Contarina moringae*
    c. *Gitona distigma*
    d. *Euptrote mollifera*
35. *Presence of ribbons of wood chips, frass and silken threads over moringa bark surface is due to* [ ]
    a. *Gitona distigma*
    b. *Indarbella tetraonnis*
    c. *Euptrote mollifera*
    d. *Noorda blitealis*

COLOUR PLATES

PESTS OF BRINJAL

Brinjal Shoot and fruit borer, *Leucinodes orbanalis*

Spotted beetle. *Hemosepilachna vigintioctopunctata*

Whitefly, ***Bemisia tabaci***

Brown leafhopper, ***Cestius phycitis***

Ash weevil, ***Myllocerus subfasciatus***

Brinjal lacewing bug, ***Urentius hystricellus***

Leafhopper, ***Amrasca devastans***

Red Spider Mite, ***Tetranychus urticae***

Aphids, ***Aphis gossypii***

PESTS OF BHENDI

Bhendi Shoot and fruit borer, *Earias vittella & Einsulana*

White fly, *Bemesia tabaci*

Leaf Hopper, *Amrasca devastanas*

Red cotton bug, *Dysdercus cingulatus*

PESTS OF TOMATO

Tomato fruit borer, ***Helicoverpa armigera***

Serpentine leaf miner, *Liriomyza trifolii*

Tomato pin worm, *Tuta absoluta*

Tomato whitefly, *Bemisia tabaci*

Tomato Thrips, *Frankliniella occidentalis*

PESTS OF CABBAGE & CAULIFLOWER

Diamond back moth, *Plutella xylostella*

Cabbage looper, ***Trichoplusia ni***

Cabbage Painted bug, ***Bagrada cruciferarum***

Cabbage aphid, ***Brevicornye brassicae***

Cabbage Head borer, ***Hellula undalis***

Cabbage butterfly, ***Pieris brassicae***

Cabbage web worm, ***Crocidolomia binotalis***

PESTS OF CUCURBITS

Cucurbit Fruit fly, *Bactrocera cucurbitae*

Pumpkin Leaf Caterpillar, *Diaphania indica*

Serpentine Leaf Miner, *Liriomyza trifolii*

Pumpkin Beteles,

***Raphidopalpa foveicollis, Aulacophora cincta* (*Grey with black border*)**
A. lewisii (Blue beetle)

whitefly, *Bemisia tabaci*

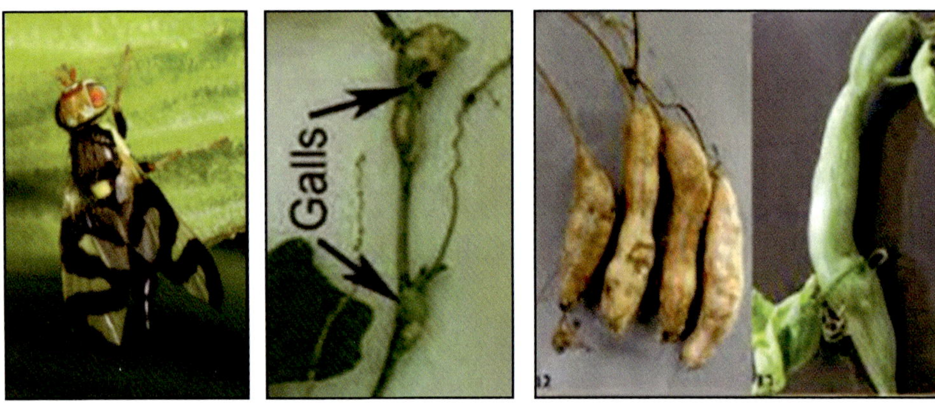

Coccinia gall fly, *Neolasioptera falcata*

Snake gourd Semilooper, Plusia peponis

PESTS OF CHILLI

Chilli thrips. *Scirtothrips dorsalis*

Chilli mite, *Polyphagotarsonemus latus*

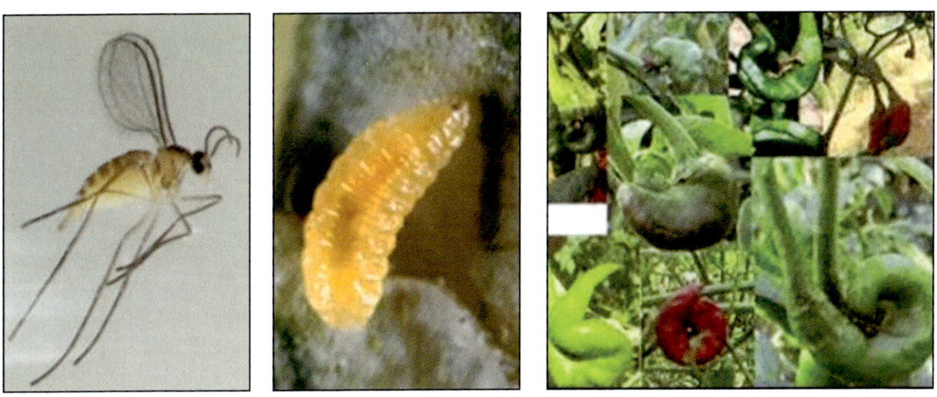

Chilli blossom midge, *Aspondylia capsici*

Tomato fruit borer, ***Helicoverpa arimegra***

Tobacco caterpillar, ***Spodoptera litira***

Sunhemp hairy caterpillar, Utethesia pulchella

Black Thrips, ***Thrips parvispinous***

PESTS OF MORINGA

Moringa hairy caterpillar, *Eupterote mollifera*

Moringa bud worm, *Noorda moringae*

Moringa pod fly, *Gitona distigma*

Bark eating caterpillar, *Indarbela tetraonis*

PESTS OF LEAFY VEGETABLE CROPS

Amaranthus weevil, *Hypolixus truncatulus*

Amaranthus leaf caterpillar, *Hymenia recurvalis*

Beet Army worm, *Spodoptera exigue*

Spinach leaf miner, *Pegomya hyoscyami*

INDEX